I0038101

DISTRIBUTED COMPUTER CONTROL SYSTEMS 2000
(DCCS 2000)

*A Proceedings volume from the 16th IFAC Workshop,
Sydney, Australia, 29 November – 1 December 2000*

Edited by

A. SOWMYA
*School of Computer Science and Engineering,
University of New South Wales, Australia*

and

J. PARK
*Department of Automation Engineering,
Inha University, Incheon, Korea*

Published for the

INTERNATIONAL FEDERATION OF AUTOMATIC CONTROL

by

PERGAMON
An Imprint of Elsevier Science

UK	Elsevier Science Ltd, The Boulevard, Langford Lane, Kidlington, Oxford, OX5 1GB, UK
USA	Elsevier Science Inc., 660 White Plains Road, Tarrytown, New York 10591-5153, USA
JAPAN	Elsevier Science Japan, Tsunashima Building Annex, 3-20-12 Yushima, Bunkyo-ku, Tokyo 113, Japan

Copyright © 2001 IFAC

All Rights Reserved. No part of this publication may be reproduced, stored in a retrieval system or transmitted in any form or by any means: electronic, electrostatic, magnetic tape, mechanical, photocopying, recording or otherwise, without permission in writing from the copyright holders.

First edition 2001

Library of Congress Cataloging in Publication Data

A catalogue record for this book is available from the Library of Congress

British Library Cataloguing in Publication Data

A catalogue record for this book is available from the British Library

ISBN 0-08-043857 1

These proceedings were reproduced from manuscripts supplied by the authors, therefore the reproduction is not completely uniform but neither the format nor the language have been changed in the interests of rapid publication. Whilst every effort is made by the publishers to see that no inaccurate or misleading data, opinion or statement appears in this publication, they wish to make it clear that the data and opinions appearing in the articles herein are the sole responsibility of the contributor concerned. Accordingly, the publisher, editors and their respective employers, officers and agents accept no responsibility or liability whatsoever for the consequences of any such inaccurate or misleading data, opinion or statement.

Printed in Great Britain

16th IFAC WORKSHOP ON DISTRIBUTED COMPUTER CONTROL SYSTEMS 2000

Sponsored by
International Federation of Automatic Control (IFAC)
Technical Committee on Distributed Computer Control Systems

Co-sponsored by
IFAC Technical Committees on:
- Artificial Intelligence in Real-Time Control
- Real-Time Software Engineering
- Safety of Computer Control Systems
- Fault Detection, Supervision and Safety of Technical Processes – SAFEPROCESS

Organised by:
School of Computer Science and Engineering, University of New South Wales, Sydney, Australia

International Programme Committee (IPC)
Jaehyun Park (Korea) (Chair)

Members:

Chan, C. (China)

Chang, N. (Korea)

Halang, W. (Germany)

Heiser, G. (Australia)

Kopetz, H. (Austria)

MacLeod, I. (South Africa)

Paoli, F. (Italy)

Parameswaran, N. (Australia)

Puschner, P. (Austria)

Ramamaritham, K. (USA)

Ramesh, S. (India)

Sahraoui, K. (France)

Sanz, R. (Spain)

Shin, H. (Korea)

Sowmya, A. (Australia)

Tovar, E. (Portugal)

Verbruggen, H. (The Netherlands)

Vingerhoeds, R. (France)

Welch, J. (USA)

Zalewski, J. (USA)

National Organizing Committee (NOC)
Sowmya, A.   (General Chair)

Members:
Hesketh, T.
Lear, J.
Roop, P.

FOREWORD

The 16th IFAC Workshop on Distributed Computer Control Systems met in Sydney, Australia, which was also the host to the 2000 Olympics and Paralympics. Development of distributed computer control systems makes demands on the disciplines of computer science and engineering as well as electrical engineering, and usually involves both hardware and software that is reliable and with assured timing properties. The DCCS series of workshops contributes to this inter-disciplinary area by bringing together academic researchers and practitioners who may normally not attend the same events.

During the past decade, DCCS has tried to focus on novel technologies for distributed control systems. This year's papers fall in the areas of classical control theory, computer architecture, real-time computer networks, and formal methods for real-time systems, besides application papers in all of these areas.

Out of 34 submitted papers, 24 were selected for presentation at the workshop. Of the 21 papers actually presented, a program subcommittee recommended 5 papers to the Editor of *Control Engineering Practice* and 1 paper to the Editor of *Automatica*, for possible publication in their journals.

This year, the workshop was co-located with PART 2000, the Seventh Australasian Conference on Parallel and Real-time Systems. This arrangement enabled us to share eminent invited speakers and also allowed DCCS participants to attend the excellent tutorials presented as a pre-conference event of PART 2000.

We would like to thank the IFAC Technical Committee on Distributed Computer Control Systems for their support in ensuring the continuity of this long-running series of workshops; IFAC and its other Technical Committees for their co-sponsorship of the event; the International Program Committee for their work in selecting the papers; and our sponsors who made this event possible.

Arcot Sowmya and Jaehyun Park

Sydney
Dec 12, 2000

CONTENTS

REAL-TIME NETWORKS - II

REAL-TIME NETWORKS –III

APPLICATIONS - II

Copyright © IFAC Distributed Computer Control Systems,
Sydney, Australia, 2000

AUTOMATION FOR SHAPE CONTROL IN STAINLESS COLD ROLLING MILL

YoneGi Hur*, **DaeKeun Rhee ****

Instrumentation and Control Research Group, Technical Research Lab.,

POSCO

1, Koedong-dong, Nam-ku, Pohang-Shi, Kyungbuk. South Korea; 790-785

*Phone: +82-54-220-6158, e-mail: tomashur@posco.co.kr

**Phone: +82-54-220-6297, e-mail: param3@posco.co.kr

Abstract: The shape of cold strip for the stainless process has been become issue in quality recently, and hence POSCO developed an automatic control system for strip shape in the sendzimir mill. The strip shape is controlled by As_U roll and first intermediate roll. A shape control system is applied to real plant. The actual shape is recognized by neural network and the fuzzy controller controls positions of actuators. The experiments are made on line and the control performance shows very efficient for the profile tracking, shape symmetry and fluctuation of shape. *Copyright© 2000IFAC*

Keywords: Shape, Fuzzy control, Neural networks, Steel industry

1. INTRODUCTION

Sendzimir mill, in stainless cold rolling process of POSCO (Pohang Iron & Steel Co., Ltd), consists of ten rolls on up and down side each. That has played main role in shape control of the stainless cold strip manually since 1990. Actuators are As_U roll, which is controlled by 8 saddles and first intermediate roll (IMR), which is driven by two pairs of rolls. The responses of actuators are not so fast since those are solenoid type; therefore operators are constrained to control these actuators precisely. What is handled in this paper is, an automatic shape controller, which is developed to minimize the fluctuation of shape deviation and to get symmetric shape. The stainless strip, one of cold-rolled product, is highly value-added product. The productivity depends on rolling speed. Improvement of rolling speed is due to the quality of strip shape so that an efficient shape control can give productivity increase in rolling line.
In the control system, the neural network generates symmetric component from real shape of strip and the fuzzy logic is applied to the shape controller, which has inputs, such as shape from measuring roll, thickness, rolling speed, positions of actuators, material type, strip width and tension and so on. That gives control outputs, which are positions of 8 saddles for As-U roll and shift values of two upper/lower first intermediate roll. The shape controller has been applied to real plant and its performance is shown very efficient. The new shape controller is very helpful to stabilize operation through automation for shape control in the sendzimir mill and improvement of shape quality.

2. SHAPE CONTROL SYSTEM FOR THE STAINLESS PROCESS

Sendzimir mill, the stainless cold rolling plant in POSCO, was established in 1990. Sendzimir mill is one stand, which consists of 20 rolls as shown in Fig. 1. As for shape meter, the measuring roll is divided into 32 zones at 52-mm intervals across strip width on both sides. The strip shape is measured by an outward measuring roll. As_U roll consists of 8 saddles, which are controlled vertically.

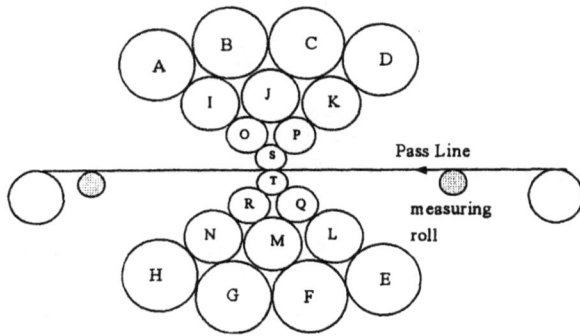

Fig. 1: A sendzimir mill

Table 1. The name of roll

Name	notation
As_U roll	B, C
Backup roll	A,D,E,F,G,H
2nd IMR (idle)	J, M
2nd IMR (drive)	I,K,L,N
1st IMR	O,P,Q,R
Work roll	S,T

The rack lies between roll B and roll C and moves up and down. The saddle, which is in the roll B (or roll C), is situated on the two backing bearings. The backing bearings are located on the second intermediate roll. If the rack moves down, the saddle of the roll B turns on its axis in the clockwise direction, but the saddle of the roll C goes round in the counterclockwise rotation. That means the saddle acts on force to the backing bearing, and then the backing bearing transfers force to the second intermediate roll. The saddle affects on rotary force to the As_U roll eccentric ring. The As-U roll acts on eccentric force to the shaft. It means force on the strip is reduced at that time (see Fig. 2).

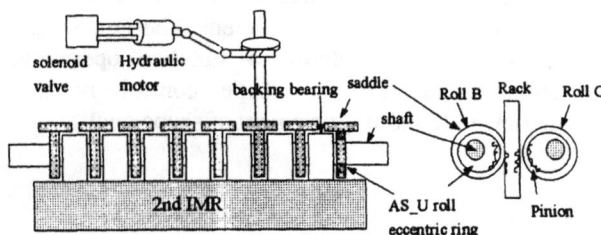

Fig. 2: The structure of As_U roll

The first intermediate rolls, which are controlled horizontally, consist of two pairs of rolls up and down. The first intermediate roll affects edge wave of strip shape since it has taper in the end of its edge side (see Fig. 3). The actuators are As_U roll (rolls B and C) and two pairs of first intermediate rolls (rolls O, P, Q and R) for shape control.

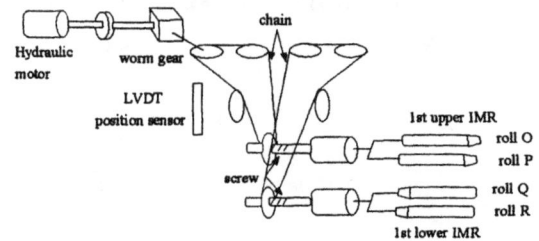

Fig. 3: The structure of first intermediate rolls

A conventional control method is that after shape pattern is classified by neural network as definition of some basic patterns, a fuzzy controller decides control outputs as much as amount belonging to those patterns "This method was developed by Katayama, *et al.* (1993), and by Hattori, *et al.* (1992) also". However it is not so easy to keep up with various shape patterns, and deduce correlation between shape variation and operation of actuators. The shape control in the stainless process is different tendency to control method under initial shape, thickness and material type of strip "This work was tested by Hur and Rhee (1999)". The fuzzy control method is proposed in this paper. It is difficult to make a control model for real process. If manual operation can be automated, it'll be possible to increase productivity and to get operator's knowledge "This is cited in the paper written by Hattori, *et al.* (1992)". For the above reason, the fuzzy control shows useful tool in various manual way of operators and situation of work.

3. THE HARDWARE CONFIGURATION OF THE SHAPE CONTROL SYSTEM

The hardware of the new shape control system composes mainly of input/output conditioner board, real-time system and MMI (Man-Machine Interface) system. The input/output conditioner board plays a role in electric or mechanic separation within a period of communicating signals inputted from sensors to the input/output board. A real-time system is central process part of the shape controller, which consists of 2 CPU (Central Process Unit), 3 analog input/output boards, and 5 digital input/output boards. One of CPU manages sequence handling and data of sensors. The other operates shape control algorithm. MMI system displays data of measuring process values and control values from CPU.

The shape controller shares input/output signals with the existing system to give process stability.

Some data are analog type, such as tension, rolling speed, shape and positions of As_U roll and first intermediate roll and so on. Some signals are digital type; for example, thickness, width, manual commands of As_U roll and first intermediate roll, limit switch, select switch between the existing system and the new system, select switch between manual and automatic mode for the new controller, signal of new strip loaded and system status and so forth. Target shape and its scale, total pass number and material type, are serial communicative type data. Data, which are mentioned above, are connected in parallel in the existing signal line. The shape of strip is inputted from an analog output board of the existing system to an analog input board of the new system. Operator selects output value between the existing system and the new shape controller by using relay. The interface between two systems is shown in Fig. 4.

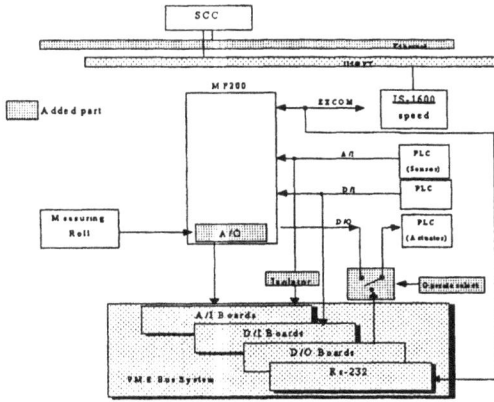

Fig. 4: The interface of a new controller

4. A SHAPE CONTROL METHOD

4.1 Recognition of shape pattern using artificial neural network

Several methods are considered for shape recognition from raw data with respect to strip width, which are to use raw data of shape directly or to do curve fitting and finally to extract symmetric component of shape using neural network. In this paper, we use some part of raw shape and symmetric shape. Using symmetric component of shape means force on strip is distributed symmetrically from the center of strip width.

Three typical shape patterns were used in the neural network, which are 2nd-order curve, 4th-order curve and 6th-order curve. All patterns have symmetric form on y-axis and the real shape can be described as various combinations of these patterns. These patterns are shown in Fig. 5. The 2nd-order curve represents center wave and edge wave and the 4th-order curve shows center wave, quarter wave, and edge wave. The 6th-order curve normally occurs on the strip width over four feet.

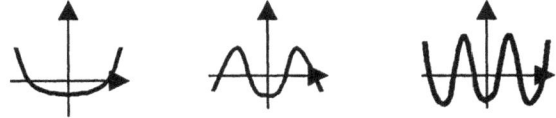

Fig. 5: Typical symmetric shape patterns

The structure of neural network is multi-layer perceptron with the learning method using back-propagation with momentum. The multi-layer perceptron has input layer, hidden layer and output layer. The input layer is composed of 32 nodes. Raw shape data are inputted into this layer. The hidden layer has 15 nodes. The output layer has 3 nodes and each node shows degree of 3 basic symmetric patterns as result of pattern classification. An example of shape recognition is shown in Fig. 6, in which the bar graph means real shape and the line graph is output of the neural network. The symmetric components of 2nd, 4th and 6th order shape are extracted from Fig. 6 and shown in Fig. 7, Fig. 8 and Fig. 9 respectively. The neural network combines these patterns to give shape for the stainless strip.

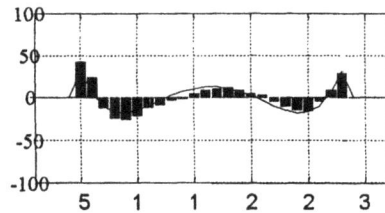

Fig. 6: Result of pattern recognition

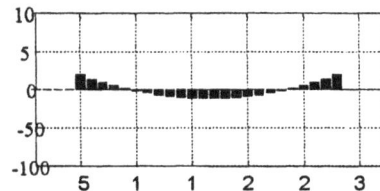

Fig. 7: 2nd-order shape pattern

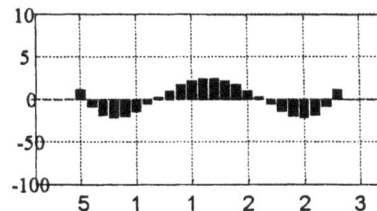

Fig. 8: 4th-order shape pattern

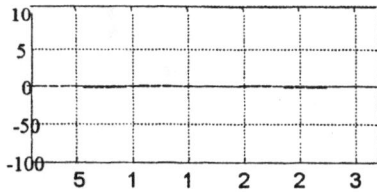

Fig. 9: 6th-order shape pattern

4.2 A shape control using fuzzy logic

A control method with As_U roll is shown in Fig. 10. After inputting positions of As_U roll and first intermediate roll, the shape controller initializes some parameters in initialized routine. That calculates DC (Direct Current) level from positions of 8 saddles and initializes gap limit between adjacent saddles and scale of control value. If manual intervention is checked, then controller outputs manual command. If the control system is in automatic mode, then the controller checks condition of rolling speed. If rolling speed is above 30 mpm, then the fuzzy control routine for As_U roll is activated. If that is above 150 mpm, then the fuzzy control routine for first intermediate roll is operated automatically.

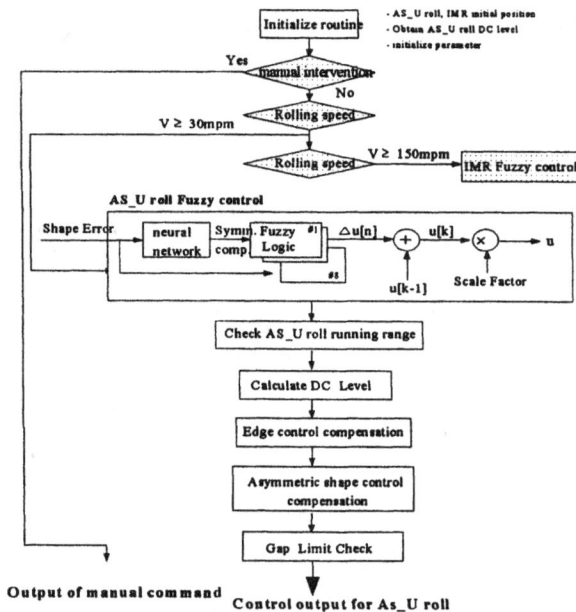

Fig. 10: The flowchart of shape control for As_U roll

The inputs of the fuzzy controller are shape error, which is the difference between the target shape and the measured shape, and its neural network output; symmetric shape. The fuzzy controller for As_U roll is 4 inputs and 1 output controller. The raw shape has 32 values because that the measuring roll is divided into 32 zones.

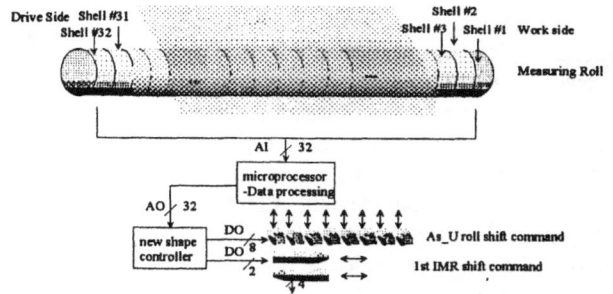

Fig. 11: Outline of the new shape controller

The outline of the new shape controller is shown in Fig. 11. The raw shape data are inputted into data processing unit from the measuring roll and then converted into analog output (AO) form. The AO data are inputted also into the new shape controller. The controller outputs As_U roll shift command and first intermediate roll shift command. The notation of "/" means the number of data.

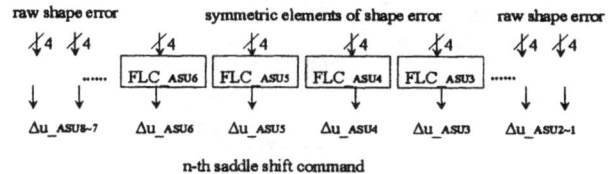

Fig. 12: Input/output of the fuzzy shape controller

The input and output of the fuzzy shape controller are shown in Fig. 12. The inputs of the fuzzy controller are the raw shape error and symmetric elements of shape error, which are extracted from the neural network. The output of the controller is shift command of n-th saddle for the As_U roll each. The 8 fuzzy controllers are used for 8 saddles respectively. The input parameter of the fuzzy controller is decided by strip width. As an example, the method of the fuzzy shape control is figured in Fig. 13, showing how to handle the center wave of the total shape.

If strip width is about 4 feet, the shape of the strip between 13th shell and 20th shell of the measuring roll is regarded as the center wave. Fig. 13 shows how to get control output; Δu_{ASU4}. That is shift value of the 4th saddle. After the shape error is fuzzified, the fuzzified shape error is composed with rules. If the shape error[13] = 1.0, the shape error[14] = 15.0, the shape error[15] = 16.0 and the shape error[16] = 16.5, and then the following rules are related to this case.

Rules:
If PV[13] = ZO & PV[14] = PS & PV[15] = PB & PV[16] = PB, Then DU = NB.

4

If PV[13] = ZO & PV[14] = PS & PV[15] = PS & PV[16] = PB, Then DU = NM.
If PV[13] = ZO & PV[14] = PS & PV[15] = PS & PV[16] = PS, Then DU = NS.

....

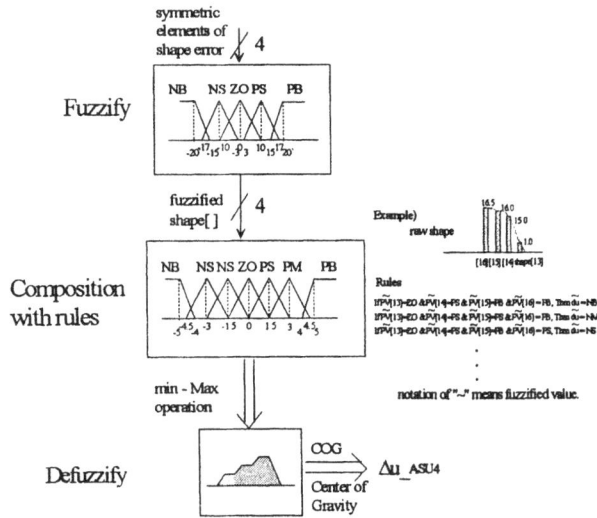

Fig. 13: The method of the fuzzy shape control

The notation is defined as follows.
[N] is N-th shell number in the measuring roll.
PV is the fuzzified shape error.
DU is the fuzzified control output.
NB is Negative Big. NM is Negative Medium. NS is Negative Small. ZO is Zero. PS is Positive Small. PM is Positive Medium. PB is Positive Big.

After composition, the membership values of the fuzzified shape errors pass through min-max operation. The fuzzified control outputs pass through defuzzification by COG (Center of Gravity) method. As a result, the fuzzy shape controller outputs shift value of the 4th saddle for As-U roll.
In above process, output of the fuzzy controller is derivative control value (Δ u_ASU). The control value u [k] is obtained as follows.

$$u\_ASU [k] = u\_ASU [k-1] + \Delta u\_ASU [k] * ScaleTc$$

where $1 \leq k \leq n$, n is integer.

The notation of ScaleTc is the scale of the control value, which is depended upon acceleration or deceleration of rolling speed, material type and output thickness of strip.

For the fuzzy control on the first intermediate roll, the shape error and the raw shape are inputted into the fuzzy controller. The two controllers operate upper and lower first intermediate roll respectively. The control value u_IMR [k] is obtained as follows.

$$u\_IMR [k] = u\_IMR [k-1] + \Delta u\_IMR [k] * ScaleIMR$$

where $1 \leq k \leq n$, n is integer.

The notation of ScaleIMR is the scale of the control value, which is depended on above same condition, but these are different.

MMI system monitors result of the shape control on line and stores shape data for analyzing the state of the process. That has several functions, such as control monitoring, 3-dimensional shape display for total pass, typical shape display for each pass, 3-dimensional shape display for each pass and I-unit display for each pass and so on. 3-dimensional shape for each pass is shown in Fig. 14.

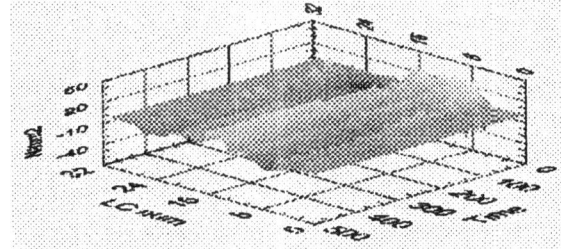

Fig. 14: 3-dimensional shape for each pass

5. TEST RESULTS

To verify performance of the new shape controller and shape control method, the test has been done for the sendzimir mill on line. The controller has been performed with respect to various material type, thickness and strip width. The test condition is that input thickness is 2.95 mm, output thickness is 1.224 mm, strip width is 1242 mm and total pass is 5. A dominant shape of 4th pass is shown in Fig. 15. It shows good shape error and symmetry. The thick line is feedback shape about strip width direction and the thin line is target shape at this pass. Y-axis is value of the load cell for each zone of the measuring roll and the unit is N/mm^2.

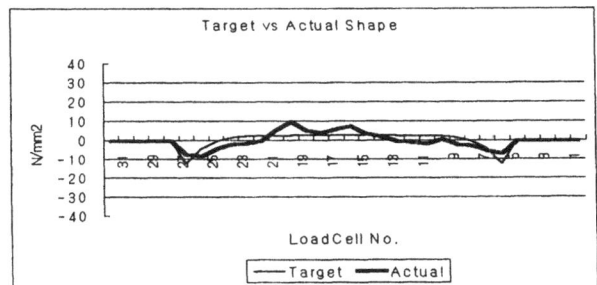

Fig. 15: Typical shape about strip width direction

The I-unit of the automatic control and the manual control is shown in Fig. 16. 1 I-unit is approximately 2.07 N/mm^2. The I-unit of the new shape control is smaller than that of the manual control. That means the shape error of the new control is stable over all passes.

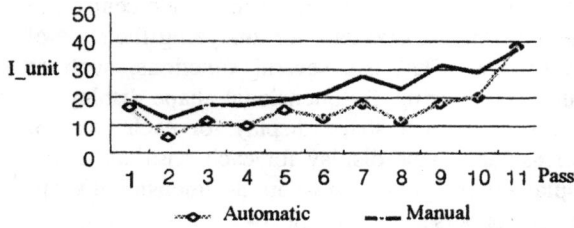

Fig. 16: I-unit of the automatic control and the manual control

6. CONCLUSIONS

This paper proposed an automatic control method for strip shape using neural network and fuzzy logic in the sendzimir mill; the stainless cold rolling mill. The shape controller has manual mode and automatic mode, which can do manual intervention as well. The new shape controller shows better performance than the previous controller does. A method of shape control is that after recognizing the symmetric component of strip shape using neural network, the shape controller outputs shift position of actuator using fuzzy logic. The experiments have been done on line for the sendzimir mill and the performance of the proposed method was verified for shape error and symmetry. Moreover, this can be obtained excellent quality and high productivity of strip as results.

REFERENCES

Katayama, Y., M. Nakajima and Y. Morooka (1993-4). Neuro-Fuzzy application to Flatness control for Sendzimir Mill. *Journal of the JSTP*, **vol. 34**, no.387, p.441- 415

Hattori, S., M. Nakajima and Y. Katayama (1992). Fuzzy Control Algorithm and Neural Networks for Flatness Control of a Cold Rolling Process. *Hitachi Review*, **vol. 41**, no.1, p.31- 38

Hur, Y.G. and D.K. Rhee (1999). Automatic Flatness Control using Artificial Neural Network and Fuzzy Technology in the STS no.1 Sendzimir Mill. *POSCO technical report*, **vol.4**, no.2, p.128-135

Copyright © IFAC Distributed Computer Control Systems,
Sydney, Australia, 2000

INTERNAL MODEL CONTROL USING RECURRENT NEURAL NETWORKS FOR NONLINEAR DYNAMIC SYSTEMS

Yan Li, David Powers and Peng Wen

School of Informatics and Engineering
The Flinders University of South Australia
GPO Box 2100, Adelaide, SA 5001, Australia

Abstract: This paper presents a control method using Recurrent Neural Networks (RNNs) in an Internal Model Control (IMC) framwork, and demonstrates their effectiveness of modelling and control for nonlinear dynamic systems. Unlike existing neural IMC design methods, the proposed control scheme consists of two stages. The first stage is that using two same structure of RNNs through iterative learning techniques to obtain a desired control signal to the unknown nonlinear systems. Then a RNN is used to generate the desired control signal within IMC structure. The algorithm for the RNNs is a real time iterative learning algorithm based on two-dimensional (2D) system theory. The simulation results demonstrate the proposed control method can drive unknown systems to follow the desired trajectories very well. *Copyright ©2000 IFAC*

Keywords: Nonlinear Control Systems, Neural Networks, Nonlinear Models,Two-dimensional systems, Multi-input/multi-output

1. INTRODUCTION

Much success has been achieved in the use of RNNs for identification and control in recent years. Many publications (Narendra and Parthasarthy, 1991; Chow and Fang, 1999; Wen *et al.*, 1996), have proved that RNNs are more powerful for nonlinear dynamic system since the architectures of RNNs themselves are presented by nonlinear dynamic system. Many approaches to train RNNs to handle time-varying input/output have been suggested or investigated by control researchers. Among those, Chow and Fang (Chow and Fang, 1999) developed a real-time iterative learning algorithm which derived by means of two-dimensional(2D) system. The algorithm is different from conventional algorithms that employ the steepest optimisation to minimise a cost function. An RNN using the algorithm can approximate any trajectory with time-varying weights. With their ability to deal effectively with time-varying

input/output, recurrent neural networks are attractive for modelling and adaptive control design.

The iterative learning algorithm is employed to train our RNNs in this paper, and internal model control (IMC) is used to provide a general framework for nonlinear system control. IMC is significant (Li *et al.*, 1996) because the stability and robustness properties of the structure can be analysed and manipulated in a transparent manner, especially for non-linear systems. The proposed control scheme consists of two stages. The first stage uses two identical RNN structures through iterative learning techniques to obtain a "desired" control signal to the unknown nonlinear systems. After the desired control signal is obtained, One of the RNNs is used to generate the desired control signal within an internal model control structure. The simulation results in this paper demonstrate the control method can drive unknown systems to follow the desired trajectories very well.

2. THE ALGORITHM

The architecture of the RNNs used in this paper is fully connected. Nodes in the RNNs are generally classified into three categories (instead of layers): input, output and hidden nodes. In this paper, we use the term "processing nodes" to represent all the output and hidden nodes. There are two sets of synaptic connections in RNNs. The first set of connection links between the input and the processing nodes. Their weights constitute the inter-weights matrix $\mathbf{W_2} = \{w_{ij}\}$. The weight $w_{ij}(t)$ $\forall i \in \mathbf{U}$ and $j \in \mathbf{I}$ (where $\mathbf{U}$ and $\mathbf{I}$ are the set of processing and input nodes, respectively) denotes the strength of the connection from the jth input node to the ith processing node, at time t. The second set of connections form the feedback paths. Therefore, each processing node is connected to all other processing nodes, including itself. Their weights constitute the intra-weight matrix $\mathbf{W_1} = w_{ij}^*$. Similarly, $w_{ij}^*(t)$ denotes the strength of the connection from the jth processing node to the ith processing node ($\forall i, j \in \mathbf{U}$), at time t. Figure 1 shows the topology of RNNs.

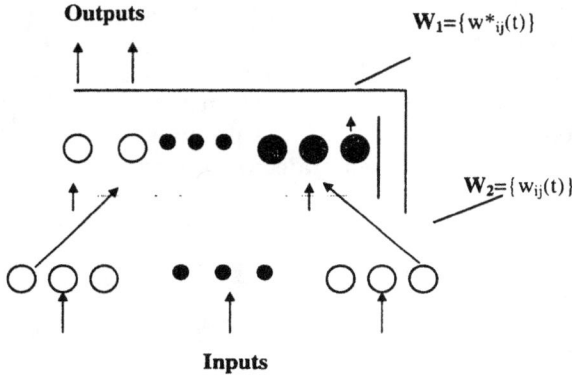

Outputs $\mathbf{W_1} = \{w^*_{ij}(t)\}$

$\mathbf{W_2} = \{w_{ij}(t)\}$

Inputs

Fig. 1. The topology of RNNs

Let $\{y(t)\}$ denote the outputs of the processing nodes and $\{u(t)\}$ denote the external inputs. Then, their state-space nonlinear dynamics of the RNN is presented in the following matrix form:

$$\mathbf{y(t+1)} = f[\mathbf{W_1(t)y(t)} + \mathbf{W_2(t)u(t)}] \quad (1)$$

with initial value $\mathbf{y(0)} = y_0$.
where

$$\mathbf{y}(t) = (y_1(t), y_2(t), ..., y_n(t))^T \in \mathbf{R}^n,$$

$$\mathbf{u}(t) = (u_1(t), u_2(t), ..., u_m(t))^T \in \mathbf{R}^m,$$

$$\mathbf{W_1(t)} \in \mathbf{R}^{nXn},$$

$$\mathbf{W_2(t)} \in \mathbf{R}^{nXn}$$

$\mathbf{f}(.)$ is a vector of a nonlinear activation function,

$$\mathbf{f}(.) = (f_1(.), f_2(.), ..., f_n(.))^T,$$

$$\mathbf{f_i}(.) = f(\mathbf{h_i}), i = 1, 2, ..., n$$

$$(h_1, h_2, ..., h_n)^T = \mathbf{W_1(t)y(t)} + \mathbf{W_2(t)u(t)}.$$

The main objective of the learning algorithm is to minimize the error function

$$\mathbf{E(t)} = \frac{1}{2} \sum_{k \in \mathbf{T(t)}} [e(t)]^2$$

$$= \frac{1}{2} \sum_{k \in \mathbf{T(t)}} [d_k(t) - y_k(t)]^2$$

$$(2)$$

through updating the weight matrices w_{ij} and w_{ij}^*. Here, $\mathbf{E(t)}$ denotes the network error at time t. The term $e_k(t)$ represents the error between the desired output $d_k(t)$ and actual $y_k(t)$, where k belongs to the set $\mathbf{T(t)}$ of the output nodes with teaching status, at time t. $e_k(t) = 0$ when k is a hidden nodes.

In this paper, we use Chow and Fang's (Chow and Fang, 1999) iterative learning algorithm based on 2-D system theory for training RNNs. According to the idea of real-time iterative learning, an algorithm based on two-dimensional expression updates the connection weights to drive the output of the network to track the desired output within a required tolerance. The learning rule can be expressed as follows:

$$\mathbf{\Delta W(t, l)} = \mathbf{C^{-1}(t)} \left[\mathbf{d(t+1)} - \mathbf{y(t+1, l-1)}\right]$$
$$\cdot \left[\mathbf{x(t)^T x(t)}\right]^{-1} \mathbf{x(t)}^T$$

$$= \mathbf{C^{-1} e(t+1, l-1)} \left[\mathbf{x^T(t)x(t)}^{-1}\right] \mathbf{x(t)}^T$$

$$(3)$$

where
$$\mathbf{C^{-1}(t)} = [diag(f'(\xi), f'(\xi), ..., f'(\xi))]^{-1},$$

$$\xi = (\xi, \xi, ..., \xi)^T = \mathbf{W(t, l-1)x(t)}.$$
$$\mathbf{x(t)} = \begin{pmatrix} \mathbf{y(t)} \\ \mathbf{u(t)} \end{pmatrix},$$
$$\mathbf{y(t+1, l)} = f^* [\mathbf{W(t, l)x(t)}],$$

$$\mathbf{W(t, l)} = \mathbf{W(t, l-1)} + \mathbf{\Delta W(t, l)}, l = 1, 2, ..., k_t.$$
Therefore, we can obtain

$$\mathbf{y(t+1)} = \mathbf{y(t+1, k_t)}$$
$$= f^* [\mathbf{W(t, k_t)x(t)}]$$
$$= f^* [\mathbf{W(t)x(t)}]$$

$$(4)$$

It is clear that the weights of an RNN are adaptively determined under the real-time algorithm. Comparing Williams and Zipser's (Williams and Zipser, 1989) algorithm with (3), this real-time iterative learning algorithm has a dynamic learning rate of $\mathbf{C}^{-1}(\mathbf{t})\left[\mathbf{x}(\mathbf{t})^{\mathbf{T}}\mathbf{x}(\mathbf{t})\right]^{-1}$. In this paper, the algorithm is applied to model and control several type of nonlinear processes.

3. INTERNAL MODEL CONTROL USING RECURRENT NEURAL NETWORKS

In this section, internal model control scheme is employed for RNNs with the real-time algorithm. There are two stages within the control scheme according to characteristics of IMC. The first step is system identification by an RNN model and the second step is neural control design based on the model inverse approach. Figure 2 shows the structure of the internal model control using RNNs.

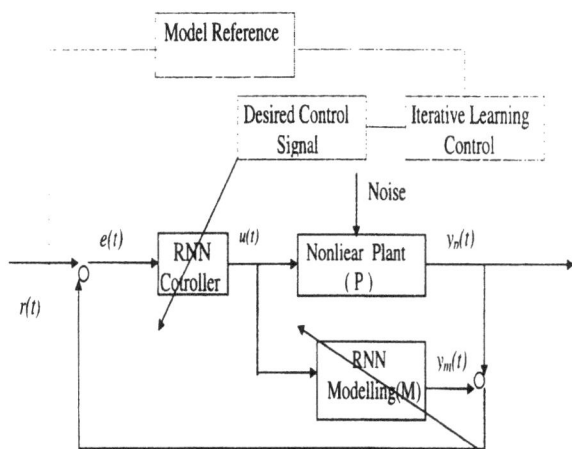

Fig. 2. The Internal Model Control Using RNNs

Stage one is an on-line identification and iterative learning control to obtain a desired control signal to the unknown nonlinear system. The idea of learning control is to utilize two RNNs, based on the same network architecture, to approximate the nonlinear system responses and to mimic the desired system response output. Then the desired control signal can be obtained by contrasting the two RNNs and the current system output and control input.

Stage 2: A recurrent neural controller is used to generate the desired control signal to the unknown nonlinear plant in an internal model control framework. The IMC structure is now well known and has been shown to underlie a number of control design techniques of apparently different origin. A unifying review of the IMC-type schemes was first presented by Garcia and Morari (Garcia and Morari, 1982). IMC has been

shown to have a number of desirable properties; a detailed analysis has been given by Morari and Zafiriov (Morari and Zafiriov, 1989).

4. THE SIMULATION RESULTS

The simulations for several unknown nonlinear dynamic systems using the proposed control scheme are given in this section. The results have demonstrated that the proposed control method can drive systems to follow the desired trajectories with high accuracy.

Example 1: Consider the unknown system which is described by the difference equation:

$$y_p(k+1) = f[y_p(k), y_p(k-1)] + u(k) \quad (5)$$

where the function

$$f[y_p(k), y_p(k-1)]$$

$$= \frac{y_p(k)y_p(k-1)(y_p(k)+2.5)}{1+y_p^2(k)+y_p^2(k-1)} \quad (6)$$

is assumed to be unknown. A reference model is described by the second-order difference equation

$$y_d(k+1) = 0.2y_d(k) + 0.2y_d(k-1) + r(k) (7)$$

where r(k) is a bounded reference input:

$$r(t) = 0.1\sin(\tfrac{2\pi t}{25})$$

According to the proposed control scheme, At first, two same RNNs are trained to get the desired control signal corresponding to the desired output. Figure 3 shows the output of the model and the output of the plant during the training. The tracking error is less than 0.001. The corresponding desired control signal is shown in Figure 4.

After the desired control signal is obtained, the RNN controller is trained on-line and the output is obtained using the internal model control structure. Figure 5 shows the control performance.

Example 2: The unknown nonlinear system is described by the following equation

$$y_p(k+1) = \frac{y_p(k)}{1+y_p^2(k)} + u^3(k) \quad (8)$$

The desired output is

$$y_m(k+1) = 0.1y_m(k) + r(k) \quad (9)$$

Assume the external input is a square wave as shown in Figure 6 (the dot line). According to the

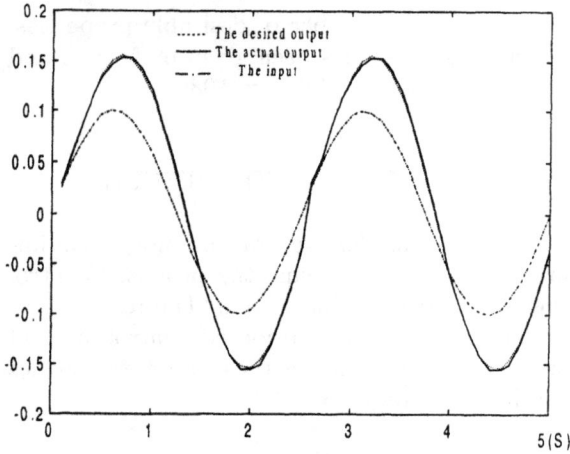

Fig. 3. The desired output and the output of the plant at stage 1 for *Example 1*

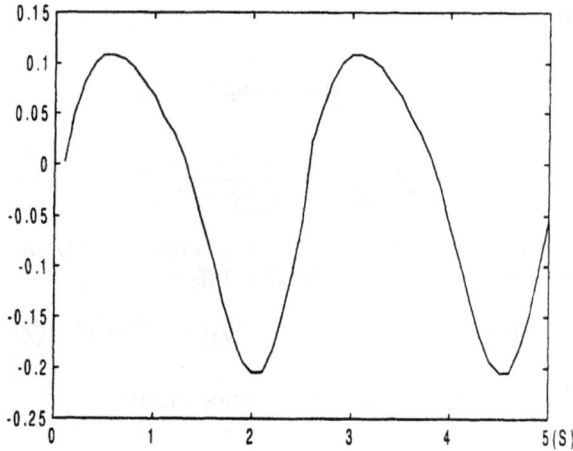

Fig. 4. The desired control signal for *Example 1*

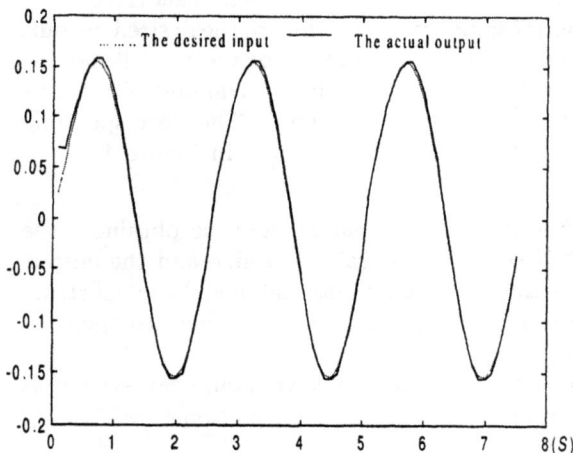

Fig. 5. The desired output and the actual output for *Example 1*

proposed control scheme, the desired control signal and the actual control signal which are shown in Figure 6 are obtained. The desired output and output of the plant obtained by internal model

control are shown in Figure 7.

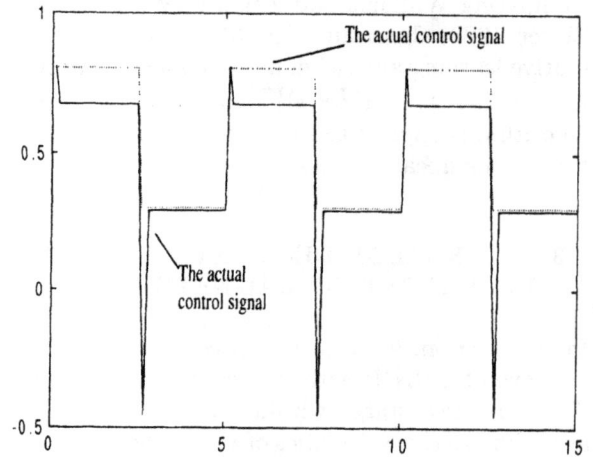

Fig. 6. The desired control signal, the actual control signal and external input for *Example 2*

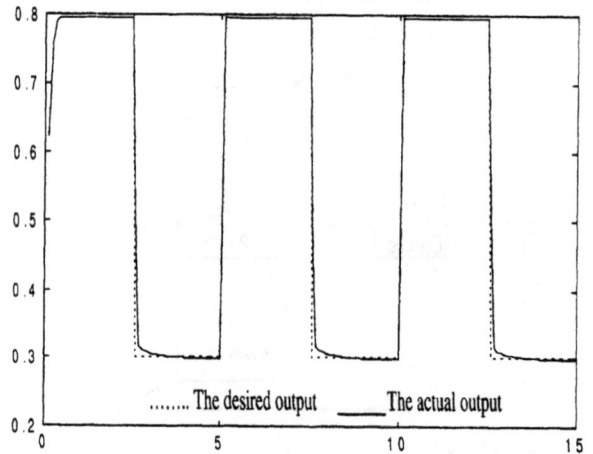

Fig. 7. The desired output and actual output for *Example 2*

Figure 8 is the performance with noise. It is observed that the outputs of the plants can track the desired trajectories very well. In *Example 1* and *2*, the structure of all the RNNs is one input, one hidden and one output node. They are all minimal networks. It is investigated that a small RNNs can implement a task as that a large or possibly infinite feedforward system does.

Example 3: The above two examples are SISO plants. In this example, an MIMO plant which is described by the following equation,

$$\begin{bmatrix} y_{p1}(k+1) \\ y_{p2}(k+1) \end{bmatrix} = \begin{bmatrix} \dfrac{y_{p1}(k)}{1+y_{p2}(k)} \\ \dfrac{y_{p1}(k)y_{p2}(k)}{1+y_{p2}^2(k)} \end{bmatrix} + \begin{bmatrix} u_1(k) \\ u_2(k) \end{bmatrix} (10)$$

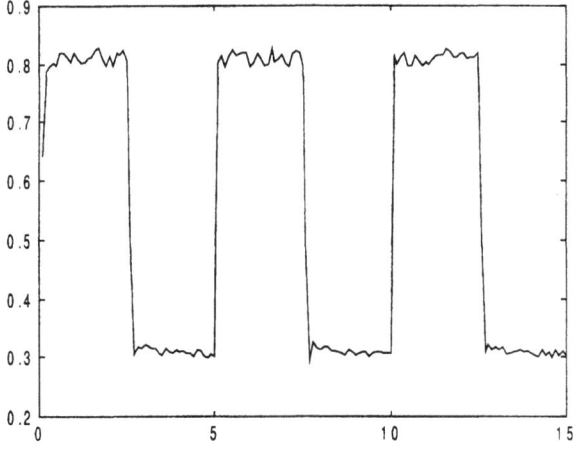

Fig. 8. The performance with noise for *Example 2*

The output of the reference model is

$$\begin{bmatrix} y_{m1}(k+1) \\ y_{m2}(k+1) \end{bmatrix} = \begin{bmatrix} 0.3 & 0.2 \\ 0.1 & 0.6 \end{bmatrix} + \begin{bmatrix} 0.3r_1(k) \\ 0.1r_2(k) \end{bmatrix} \quad (11)$$

Here, the external inputs have the form of

$$\begin{bmatrix} r_1(k) \\ r_2(k) \end{bmatrix} = \begin{bmatrix} \sin(\dfrac{2\pi k}{25}) \\ \cos(\dfrac{2\pi k}{25}) \end{bmatrix} \quad (12)$$

The desired control signals with respect to the desired outputs and the actual control signals U_{d1}, U_{d2} and U_{i1}, U_{i2} are shown in Figure 9 and 10.

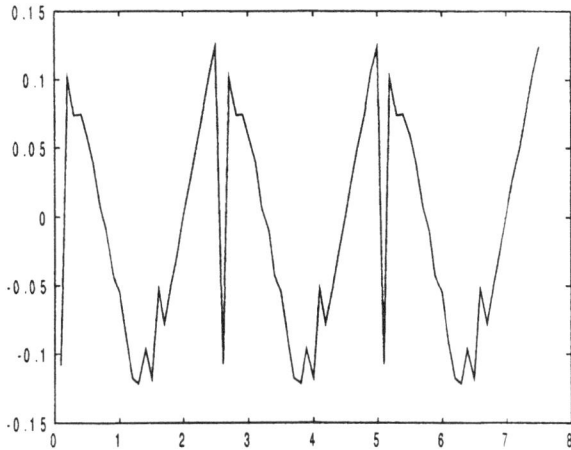

Fig. 9. The actual control signal U_{i1} and the desired control signal U_{d1} for *Example 3*

Figure 11, 12 show the final control results, two outputs Y_{p1}, Y_{p2} and the desired outputs Y_{d1}, Y_{d2}.

In this example, the structure of the two RNNs consists of two input nodes, two hidden nodes and two output nodes. It is observed that the errors between the actual outputs of the plant and the desired outputs are larger than those in the SISO

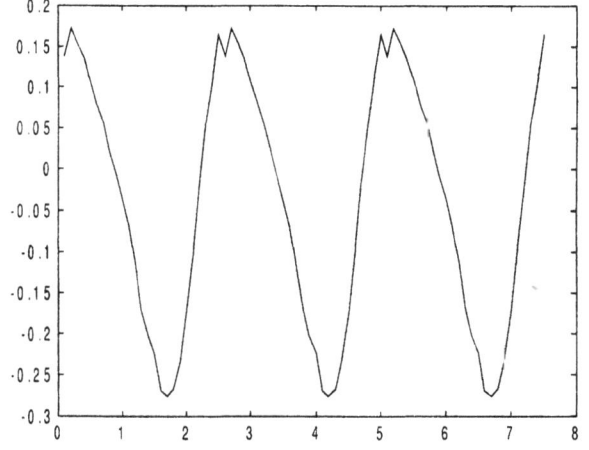

Fig. 10. The actual control signal U_{i2} and the desired control signal U_{d2} for *Example 3*

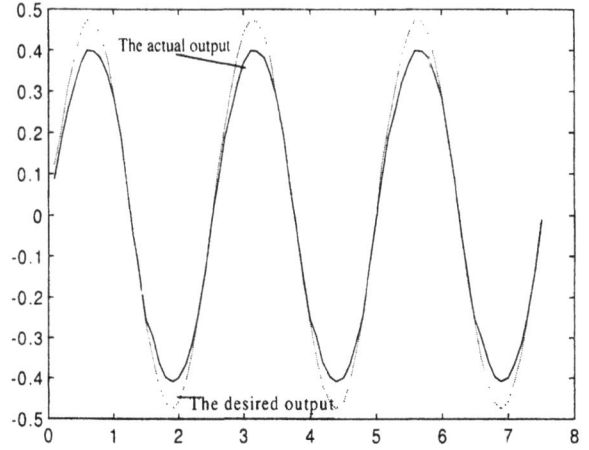

Fig. 11. The output Y_{p1} of the plant and the desired output for *Example 3*

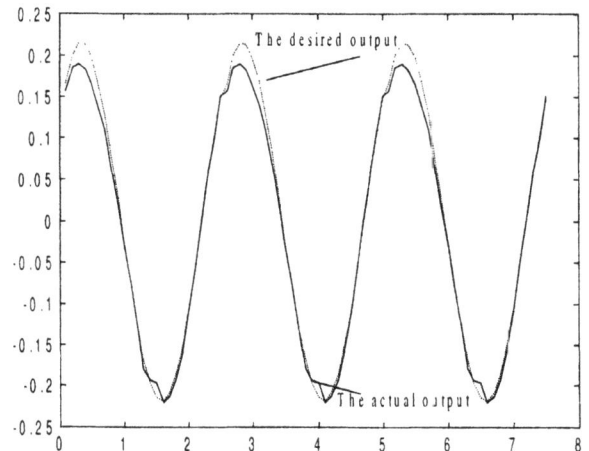

Fig. 12. The output Y_{p2} of the plant and the desired output for *Example 3*

situations. The main reason for this is that the errors between each of the two actual outputs and their corresponding desired outputs cannot reach the specified error tolerance simultaneously.

5. CONCLUSION

The paper presents a control method of using RNNs in an IMC framwork, and demonstrates their effectiveness for modelling and control of nonlinear dynamic systems. Unlike existing neural IMC design methods, the proposed control scheme consists of two stages. The first stage uses two same structure of RNNs through iterative learning techniques to obtain a desired control signal to the unknown nonlinear systems. Then one of the RNNs is used to generate the desired control signal within IMC structure. The algorithm for the RNNs is a real time iterative learning algorithm based on two-dimensional (2D) system theory. The simulation results for both the SISO and MIMO situations demonstrate the proposed control method can drive unknown systems to follow the desired trajectories very well.

6. REFERENCES

Chow, T.W.S. and Y. Fang (1999). A recurrent neural-network-based real-time learning control strategy applying to nonlinear systems with unknown dynamics. *IEEE Transactions on industrial electronics* **45(1)**, 151–161.

Garcia, C. and M. Morari (1982). Internal model control-1: A unifying review and some new results. pp. 308–326.

Li, Y., A. B. Rad and Y K Wong (1996). Model based control using artificial neural network. *Proceedings of the 1996 IEEE international symposium on intelligent control* **1**, 15–18.

Morari, M. and E. Zafiriov (1989). *Robust Process Control*. Prentice-Hall.

Narendra, K.S. and K. Parthasarthy (1991). Gradient methods of the dynamical systems containing neural networks. *IEEEE Transactions on Neural Networks* **2(2)**, 7–21.

Wen, P., C.K. Ng and Y. Li (1996). Dynamic linear square backpropagation algorithm for recurrent neural networks. *The Fourth International Conference on Control, Automation, Robotics and Vision* **1**, 3–6.

Williams, R. and D. Zipser (1989). A learning algorithm for continually running fully recurrent neural networks. *Neural Computation* **1**, 270–280.

Copyright © IFAC Distributed Computer Control Systems,
Sydney, Australia, 2000

STOCHASTIC CONTROL OF COMMUNICATION NETWORK WITH TIME VARYING DELAY

Eik-Dong Park * Uk-Youl Huh *

Dept. of Elec. Eng., Inha Univ., Inchon, 402-751, Korea
Tel : +82-32-860-7394
Fax : +82-32-863-5822
E-mail : g2001110@inhavision.inha.ac.kr
**Dept. of Elec. Eng., Inha Univ., Inchon, 402-751, Korea*
Tel : +82-32-860-7394
Fax : +82-32-864-6442
E-mail : uyhuh@inha.ac.kr

Abstract: Communication networks are shared with other users. There will inevitably be time delays in the communication net. The traffic condition in the network may introduce time-varying random delays in the control loop with on its performance and stability. Hence, the control must be designed to compensate for delays. In this paper, modelling integrated control and communication systems has been shown by inducing random communication delays. And numerical test for zero-state stability is shown. *Copyright © 2000 IFAC*

Keywords: delay, stochastic control, DCS, real-time system

1. INTRODUCTION

Many real-time systems are implemented as distributed control systems, where the control loops are closed over a communication network or a field bus. There will inevitably be time delays in the communication net. As long as the sampling periods are long compared with these delays there is no need to consider the influence of the delays. Complex dynamical processes like advanced aircraft, spacecraft, and autonomous manufacturing plants require high-speed, reliable communications between system components which perform a set of inter-related functions ranging from active control to information display and routine maintenance support. The system components include a number of computers, intelligent terminals, sensors and actuators, and their functions are executed in real time. The activities of systems components can be coordinated by appropriate information exchange via a multiplexed communication network to achieve a better utilization of the resources. However, the network introduces delays in addition to the sampling delay that is prevalent in all digital control systems. The network-induced delay are time-varying and possibly stochastic, and are dependent on the intensity, probability distribution, dynamics of the traffic as well as missynchronization between control system components and noise in the communication medium. As the demand on the control system increases it will be more and more important to take the delays into account in the analysis and the design of the control system. While inaccuracies, disturbances, etc., have been extensively studied in the control literature the timing problems in real-time systems have just recently attracted attention, and in the communication literature the feedback control aspect has not been treated to any larger extent. This is thus an area where much can be gained by combining ideas from the fields control, real-time systems, and communication networks. The effect of communication delays is discussed, for

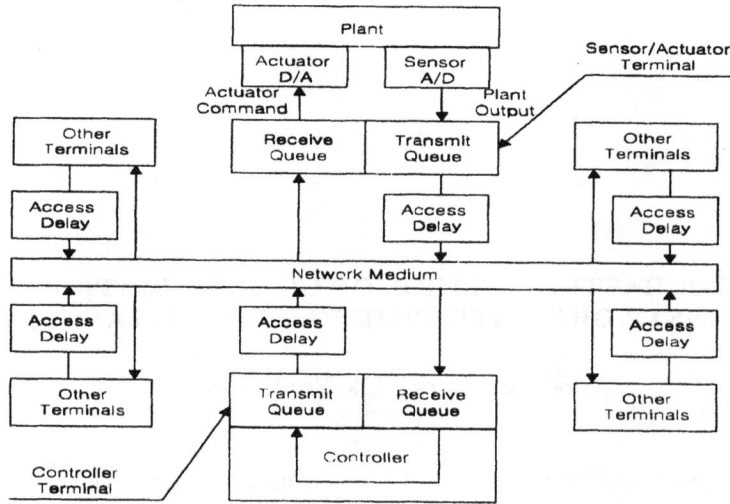

Fig. 1. Schematic diagram for the ICCS

instance, in Ray (Ray, 1989), Chan (Chan and Özgüner, 1995). Therefore, the Integrated Communication and Control System(ICCS) for these processes must be designed to compensate for these delays. The schematic diagram of an ICCS network in Fig.1 illustrates how these delays are introduced (Ray, 1988). In ICCS, a control loop is closed via the common communication channel which multiplexes digital data from the sensor to the controller and from the controller to the actuator along with the data traffic from other control loops and management functions. Furthermore, the control system components(e.g. the sensor and controller) may not be synchronized. The theoretical work on stability characterization of linear discrete–time systems with randomly time-varying delays was used for stability analysis of asynchronous multirate control systems (Ray and Halevi, 1988). Discrete time–varying state space representation is obtained. It is observed that the randomly varying structure of the state–space representation may be described as a jump random process, so that the overall system falls into the class of hybrid or jump systems (Mariton, 1990). Related results can also be found in the work of (Chizeck *et al.*, 1986) and (Birdwell and Castanon, 1986). In this paper, single communication medium and the simplest case of a synchronous single–rate feedback of a single–channel feedback system is considered.

2. GENERAL PROBLEM FORMULATION

Let us consider a realization of a control loop illustrated in Fig. 1. We are interested in the single–controller single–actuator case. Thus there is only one controller–to–actuator(C/A) and one sensor–to–controller(S/C) delay of concern. As mentioned before, traffic may be produced in

(C/A) and/or (S/C) communication line by message issued simultaneously from different control or information subsystems, which induces unpredictable delays in the loop of a control subsystem. For the general case of a dynamic controller such a control loop may be described by the following equations.

$$\left.\begin{aligned}
x_{k+1} &= Ax_k + Bu_k \\
y_k &= Cx_k \\
p_{k+1} &= Fp_k + Gw_k \\
v_k &= Hp_k + Kw_k \\
u_k &= \sum_{i=0}^{D_1} \alpha_i(k)v_{k-i} \\
w_k &= \sum_{i=0}^{D_2} \beta_i(k)y_{k-i}
\end{aligned}\right\} \quad (1)$$

where

$$k = 0, 1, 2, \ldots$$
$$A, B, C, F, G, H, K = constant$$
$$\alpha_i(k), \beta_i(k) \in \{0, 1\}(k = 0, 1, \ldots),$$
$$i = 0, 1, \ldots, D_\nu, \quad D_\nu \in N(\nu = 1, 2)$$
$$\sum_{i=0}^{D_1} \alpha_i(k) = 1, \quad \sum_{i=0}^{D_2} \beta_i(k) = 1$$

In other word, the possibility of C/A and S/P delays ranging from zero to $D_v \Delta$ t is considered ($\nu = 1, 2$). Replacing $y_k = Cx_k$ and $v_k = Hp_k + Kw_k$ in the last two relations,we obtain a more convenient form of the model(1)

$$\left.\begin{aligned}
x_{k+1} &= Ax_k + Bu_k \\
w_{k+1} &= C \sum_{i=0}^{D_2} \beta_i(k+1)x_{k+1-i} \\
p_{k+1} &= Fp_k + Gw_k \\
u_{k+1} &= H \sum_{i=0}^{D_1} \alpha_i(k+1)p_{k+1-i} \\
&\quad + K \sum_{i=0}^{D_1} \alpha_i(k+1)w_{k+1-i}
\end{aligned}\right\} \quad (2)$$

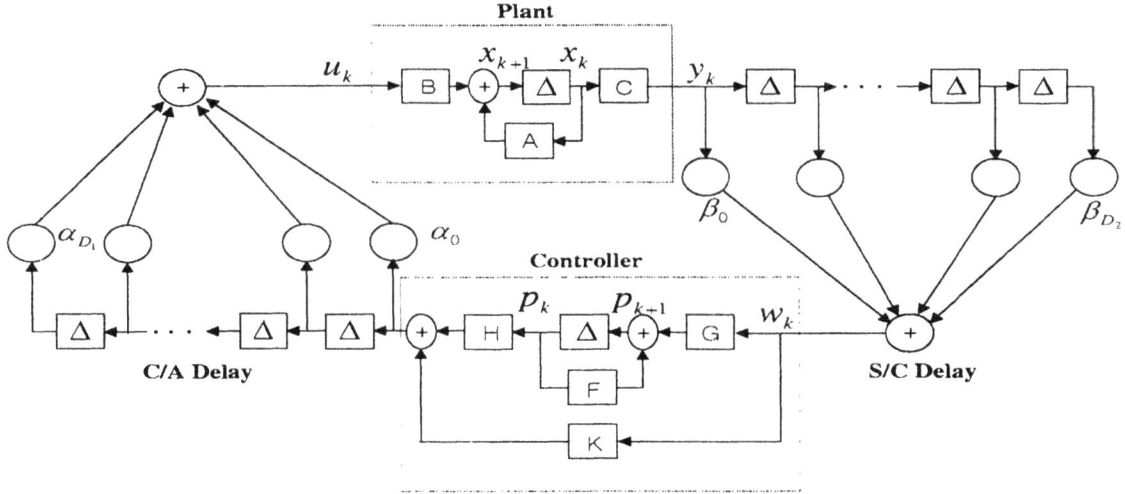

Fig. 2. Block diagram illustrating the model (1).

3. DISCRETE TIME-VARYING SYSTEM REPRESENTATION

In order to obtain a discrete time-varying state-space representation of the control system with random communication delays (1), we denote

$$
\begin{aligned}
r(k) &:= \left[x_k^T \vdots r_1^T(k) \vdots r_2^T(k) \vdots \cdots \vdots r_{D_2-1}^T(k) \right]^T \\
&:= \left[x_k^T \vdots x_{k-1}^T(k) \vdots x_{k-2}^T(k) \vdots \cdots \vdots x_{k-D_2+1}^T(k) \right]^T \\
s(k) &:= \left[w_k^T \vdots s_1^T(k) \vdots s_2^T(k) \vdots \cdots \vdots s_{D_1-1}^T(k) \right]^T \\
&:= \left[w_k^T \vdots w_{k-1}^T(k) \vdots w_{k-2}^T(k) \vdots \cdots \vdots w_{k-D_1+1}^T(k) \right]^T \\
t(k) &:= \left[p_k^T \vdots t_1^T(k) \vdots t_2^T(k) \vdots \cdots \vdots t_{D_1-1}^T(k) \right]^T \\
&:= \left[p_k^T \vdots p_{k-1}^T(k) \vdots p_{k-2}^T(k) \vdots \cdots \vdots p_{k-D_1+1}^T(k) \right]^T
\end{aligned}
\tag{3}
$$

Using (2) and (3), we are able to express u_{k+1} as a linear combination of vectors $r(k)$, $s(k)$ and $t(k)$.

$$
\begin{aligned}
u_{k+1} = R_{k+1} + S_{k+1}s(k) + T_{k+1}t(k) \\
+ \alpha_0(k+1)\beta(k+1)KCBu_k
\end{aligned}
\tag{4}
$$

where

$$
R_{k+1} := \alpha_0(k+1)[\beta_0(k+1)KCA
$$
$$
+ \beta_1(k+1)KC \vdots \beta_2(k+1)KC \vdots \cdots \vdots
$$
$$
\beta_{D_2}(k+1)KC]
$$

$$
S_{k+1} := [\alpha_0(k+1)HG + \alpha_1(k+1)K \vdots \alpha_2(k+1)K \vdots
$$
$$
\cdots \vdots \alpha_{D_1}(k+1)K]
$$

$$
T_{k+1} := [\alpha_0(k+1)HF + \alpha_1(k+1)H \vdots \alpha_2(k+1)H \vdots
$$
$$
\cdots \vdots \alpha_{D_1}(k+1)H]
$$

The recurrence equations for $r(k)$, $s(k)$ and $t(k)$ are easily obtained from their definition (3) and the first relations in (2). We use the notation

$$
\left.
\begin{aligned}
r_{k+1} &= \overline{R}r(k) + \overline{B}u(k) \\
s_{k+1} &= \overline{S}_{k+1}r(k) + \overline{I}s(k) + \overline{J}_{k+1}u(k) \\
t_{k+1} &= \overline{T}t(k) + \overline{G}s(k)
\end{aligned}
\right\}
\tag{5}
$$

Recurrence relation (4) and (5) define the state-space representation of the system (1). We choose the state vector to be

$$
z := \left[r^T(k) \vdots s^T(k) \vdots t^T(k) \vdots u_k^T \right]^T
\tag{6}
$$

To write (4) and (5) in compact form, we define

$$
H_k := \begin{bmatrix}
\overline{R} & 0 & 0 & \overline{B} \\
\overline{S}_{k+1} & \overline{I} & 0 & \overline{J}_{k+1} \\
0 & \overline{G} & \overline{T} & 0 \\
\overline{R}_{k+1} & \overline{S}_{k+1} & \overline{T}_{k+1} & \alpha_0(k+1)\beta_0(k+1)KCB
\end{bmatrix}
\tag{7}
$$

Using (6) and (7) we rewrite (4) and (5) as follows

$$
z_{k+1} = H_k z_k
\tag{8}
$$

Relation (8) is a standard discrete time-varying state-space representation of the control system (1) with dynamic controller and random multiple-time-step communication delays in the sensor and actuator channels.

It should be noted that the matrix H_k dependence on the time parameter k is only due to the presence of the binary sequence $\alpha_i(k+1)$, $\beta_i(k+1)(i,j = 0, 1, 2, \ldots, D_\nu)$. For each pair of non-zero values $\alpha_i(k+1)$, $\beta_j(k+1)$ (which corresponds to the occurrence of the C/A delay $i\Delta t$ and the S/C delay $j\Delta t$ in the communication links), we have a particular (time-invariant) system structure described by the state matrix A_{ij}. There is no more than $(D_1 + 1)(D_2 + 1)$ different system structure in the close-loop system in which the maximal possible C/A and S/C delays are equal to D_ν. We have therefore

$$H_k \in \{A_{ij}; i = 0, 1, 2, \ldots, D_1; j = 0, 1, 2, \ldots, D_2\} \tag{9}$$

Hence, the system matrix H_k can be viewed as a finite automation with $(D_1+1)(D_2+1)$ states corresponding to different system structures A_{ij} where the state transition function is defined by the sequences $\alpha_i(k+1), \beta_i(k+1)(i,j = 0,1,2,\ldots,D_\nu)$. Whenever $\alpha_i(k+1) = \alpha_i(k), \beta_i(k+1) = \beta_i(k)$, we have $H_k = H_{k-1}$ and the transition $z_{k+1} = H_{k-1}^2 z_{k-1}$ is performed as if the system was time invariant. Thus the time-varying state-space representation(8) of the system(1) may be replaced by the description of all possible time-invariant system structures A_{ij}, and the description of transitions between those structures. Such a description is sometimes called a hybrid system or a jump system.

4. STABILITY ANALYSIS WITH DYNAMIC CONTROL AND MULTIPLE-STEP DELAYS

4.1 *Reduction of the general problem to a standard form*

Indeed, the system(8) is a hybrid system with jump parameters $\alpha_i(k), \beta_i(k)$

$$M := (D_1 + 1)(D_2 + 1) \tag{10}$$

different system structures defined by the state matrices A_{ij} in(9). The matrix A_{ij} is the state of the system(8) at time k if $\alpha_i(k) = \beta_j(k) = 1$. As mentioned, the linear discrete time-varying system(8), rewritten here as

$$z_{k+1} = H_k z_k \tag{11}$$

may be viewed as a process jumping over a set of time-invariant structures A_{ij}. It is more convenient to arrange these states and the corresponding matrices A_{ij} in line, introducing the index

$$\nu := i(D_2 + 1) + j + 1$$
$$(i = 0, 1, 2, \ldots, D_1; j = 0, 1, 2, \ldots, D_2) \tag{12}$$

so that $\nu = 1, 2, \ldots, M$. In this way we are able to use an alternative notation for the matrices A_{ij}:

$$A_\nu := A_{ij}(i = 0, 1, 2, \ldots, D_1; j = 0, 1, 2, \ldots, D_2$$
$$\nu = 0, 1, 2, \ldots, M) \tag{13}$$

With each structure A_ν is associated a state S_ν of the jump process. Let us define a binary index

$$\gamma_\nu(k) := \alpha_i(k)\beta_j(k) \tag{14}$$

and observe that $\sum_{i=1}^{M} \gamma_i(k) = 1$. It follows that

$$H_k = \sum_{\nu=1}^{M} \gamma_\nu(k)A_\nu \tag{15}$$

The state of the jump process at time k is determined by the vector of M binary variables

$$\theta_k := [\gamma_1(k)\gamma_2(k) \ldots \gamma_M(k)]^T \tag{16}$$

which is, as in the previous section, assumed to be a Markov chain with finite number of states $S_\nu; \nu = 1, 2, \ldots, M$. The transition conditional probability that the process will jump at time $k = 0, 1, 2, \ldots$ from state $S_j(j = 1, 2, \ldots, M)$ to state S_j is denoted as $p_{ij}(k)$. The $M \times M$ transition probability matrix is denoted as

$$\Pi := \begin{bmatrix} p_{11}(k) & p_{12}(k) & \ldots & p_{1M}(k) \\ p_{21}(k) & p_{22}(k) & \ldots & p_{2M}(k) \\ \vdots & \vdots & & \vdots \\ p_{M1}(k) & p_{M2}(k) & \ldots & p_{MM}(k) \end{bmatrix} \tag{17}$$

The probability of finding the jump process in state $S_\nu(\nu = 1, 2, \ldots, M)$ at time $k = 0, 1, 2, \ldots$ is denoted as $p_\nu(k)$; the $M \times 1$ vector of probabilities $p_\nu(k)(\nu = 1, 2, \ldots, M)$ at time k is denoted as

$$p_k := [p_1(k)p_2(k) \ldots p_M(k)]^T \tag{18}$$

Complete description of the jump system(11) includes also the transition probabilities of Markov chain:

$$p_{k+1} = \Pi_k p_k \tag{19}$$

In shorthand notation, (15)can be written as $H_k = H_k(\theta_k)$ so that

$$\left. \begin{array}{l} \Phi(i,k) = H_{i-1}(\theta_{i-1})H_{i-2}(\theta_{i-2}) \ldots H_k(\theta_k) \\ \Phi(k,k) = I_N \\ \quad i = k+1, k+2, \ldots \end{array} \right\} \tag{20}$$

Relations (11),(19) and (20) show that the system with multiple delays(and dynamic feedback) is a straightforward extension of the system with one possible delay(and static feedback) treated.

4.2 *Stochastic stability*

Let us assume that $d_k; k = 0, 1, 2, \ldots$ is a Markov chain with two states S_1 and S_2 corresponding to the realizations $d_k=0$ and $d_k=1$ of the binary random variable d_k, and the transition probability matrix at time k given by

$$\Pi_k := \begin{bmatrix} p_{11}(k) & p_{12}(k) \\ p_{21}(k) & p_{22}(k) \end{bmatrix} \tag{21}$$

where $p_{ij}(k)$ is the conditional probability that the chain being in the state S_j at time $k-1$ will jump to the state S_i at time k. Hence the $p_{ij}(k)$ satisfy the constraints

$$0 \leq p_{ij}(k) \leq 1, p_{ij}(k) + p_{2j}(k), \ i,j = 1,2 \tag{22}$$

The probabilities that the chain will find itself in the states S_1 and S_2 at time k are denoted as $p_1(k)$ and $p_2(k) = 1 - p_1(k)$ and are determined by the equation

$$p_{k+1} = \Pi_k p_k \qquad (23)$$

where $p_k := [p_1(k) p_2(k)]^T$. We assume that $p_0 = [10]^T$. The states z_k given by

$$\left. \begin{array}{l} z_{k+1} = H_k(d_k) z_k \\ H_k(d_k) = A_0 + d_k A_c \end{array} \right\} \qquad (24)$$

A_0 and A_1 denote constant matrices obtained from H_k for $d_k = 0$ and $d_k = 1$ respectively. Hence the general time–varying state representation (8) is due to irregular switching between two system structures described by matrices A_0 and

$$A_1 := A_0 + A_c \qquad (25)$$

For any pair of symmetric positive definite matrix sequences $Q_k(0), Q_k(1)$ such that

$$\left. \begin{array}{l} 0 < c_1 \le \lambda_{min}[Q_k(0)] \le \lambda_{max}[Q_k(0)] \le c_2 < \infty \\ 0 < c_3 \le \lambda_{min}[Q_k(1)] \le \lambda_{max}[Q_k(1)] \le c_4 < \infty \end{array} \right\}$$
$$(26)$$

The system (24) is mean square exponentially stable if and only if there exist symmetric positive definite matrix sequences $P_k(0), P_k(1)$ such that

$$\left. \begin{array}{l} H_k^T(0) \left[p_{11}(k+1)P_{k+1}(0) + p_{21}(k+1)P_{k+1}(1) \right] \\ \qquad H_k(0) + Q_k(0) = P_k(0) \\ H_k^T(1) \left[p_{12}(k+1)P_{k+1}(0) + p_{22}(k+1)P_{k+1}(1) \right] \\ \qquad H_k(1) + Q_k(1) = P_k(1) \end{array} \right\}$$
$$(27)$$

where

$$\left. \begin{array}{l} 0 < c_5 \le \lambda_{min}[P_k(0)] \le \lambda_{max}[P_k(0)] \le c_6 < \infty \\ 0 < c_5 \le \lambda_{min}[P_k(1)] \le \lambda_{max}[P_k(1)] \le c_8 < \infty \end{array} \right\}$$
$$(28)$$

and $\lambda_{min}[\cdot]$ and $\lambda_{max}[\cdot]$ mean the minimal and the maximal eigenvalues respectively.// Let us consider the transition matrix defined in (20) and the sequence of symmetric positive-definite matrices $Q_k(r_k)$ uniformly bounded. Under the mean square exponential stability assumption, we are able to define the matrix

$$\overline{P}_k := \sum_{i=k}^{\infty} \Phi^T(i,k) Q_i \Phi(i,k) \qquad (29)$$

which implies that

$$\overline{P}_k = H_k^T(r_k) \overline{P}_{k+1} H_k(r_k) + Q_k(r_k) \qquad (30)$$

We introduce the matrix

$$P_k(\nu) := E(\overline{P}_k | r_k = \nu) \qquad (31)$$

Relation (30) implies that

$$E(\overline{P}_k | r_k) = H_k^T(r_k) E(\overline{P}_{k+1} | r_k) H_k(r_k) + Q_k(r_k) \qquad (32)$$

Note that

$$E(\overline{P}_{k+1} | r_k) = E \left[E(\overline{P}_{k+1} | r_k, r_{k+1}) | r_k \right] \qquad (33)$$

and that the Markov property implies that

$$E(\overline{P}_{k+1} | r_k, r_{k+1}) = E(\overline{P}_{k+1} | r_{k+1}) \qquad (34)$$

Hence

$$E(\overline{P}_{k+1} | r_k = \nu) = E \left[E(\overline{P}_{k+1} | r_k) | r_k = \nu \right] \qquad (35)$$

Therefore

$$E(\overline{P}_{k+1} | r_k = \nu) = \sum_{i=1}^{M} p_{i\nu}(k+1) E(\overline{P}_{k+1} | r_{k+1} = i)$$
$$= \sum_{i=1}^{M} p_{i\nu}(k+1) P_{k+1}(i) \qquad (36)$$

Substituting (31) and (36) in (32, we obtain the generalization of the stability equation

$$P_k(\nu) = \sum_{i=1}^{M} p_{i\nu}(k+1) A_\nu^T P_{k+1}(i) A_\nu + Q_k(\nu) \qquad (37)$$

because the state matrix $H_k(r_k)$ is equal to A_ν for $r_k = \nu$. With stability equation (37), the generalization of (27) to the multiple-delay case is straightforward.

5. NUMERICAL EXAMPLE

The design of discrete time stabilizing control of a continuous time model as following will be shown.

$$\dot{x}(t) = u(t), \; y(t) = x(t) \qquad (38)$$

The corresponding discrete time model is

$$x_{k+1} = x_k + T_s u_k, \; y_k = x_k \qquad (39)$$

where T_s is a sampling period. The binary variable d_{k-1} to indicate the presence of the communication delay of the length T_s in the kth time step(i.e. $d_{k-1} = 1$ in this case and $d_{k-1} = 0$ otherwise).

$$x_{k+1} = [1 - gT_s(1 - d_{k-1})] x_k - gT_s d_{k-1} x_{k-1} \qquad (40)$$

Let

$$\left. \begin{array}{l} a := gT_s \\ \alpha_k := \alpha(d_{k-1}) = 1 - a(1 - d_{k-1}) \\ \beta_k := \beta(d_{k-1}) = a d_{k-1} \end{array} \right\} \qquad (41)$$

and state space form

$$z_{k+1} = H_k z_k \qquad (42)$$

where

$$H_k := h(d_{k-1}) = \begin{bmatrix} \alpha_k & -\beta_k \\ 1 & 0 \end{bmatrix} \qquad (43)$$

And

$$A_0 := H(0) = \begin{bmatrix} 1-a & 0 \\ 1 & 0 \end{bmatrix}$$

$$A_1 := H(1) = \begin{bmatrix} 1 & -a \\ 1 & 0 \end{bmatrix}$$

$$A_c := A_1 - A_0 = \begin{bmatrix} a & -a \\ 1 & 0 \end{bmatrix}$$

The simplest assumption concerning random occurrence of communication delays is that they are uniformly distributed.($p_{11} = p_{12} = p_{21} = p_{22} = \frac{1}{2}$)

$$\left. \begin{aligned} A_0^T P(0) A_0 + A_0^T P(1) A_0 + 2I_2 &= 2P(0) \\ A_1^T P(0) A_1 + A_1^T P(1) A_1 + 2I_2 &= 2P(1) \end{aligned} \right\} \quad (44)$$

For a=1

$$P(0) = \begin{bmatrix} 5 & 0 \\ 0 & 1 \end{bmatrix} \qquad (45)$$

$$P(1) = \begin{bmatrix} 7 & -4 \\ -4 & 7 \end{bmatrix} \qquad (46)$$

$P(0)$ and $P(1)$ are positive definite. Therefore system (42) is mean square exponentially stable. For a=2, the system is not mean square exponentially stable.

6. CONCLUSION

This paper presents a modelling for decentralized systems with time–varying random delay feedback through a communication medium . And the necessary and sufficient condition for stability of that system is shown. Using the condition, stability of a simple numerical system is shown.

7. REFERENCES

Anderson, B. D. O. and J. B. Moore (1979). *Optimal Filtering*. Prentice-Hall.

Åström, K. J. (1970). *Introduction to Stochastic Control Theory*. Academic Press.

Birdwell, J. D. and D. A. Castanon (1986). On reliable control system design. *IEEE Trans. on Systems, Man, and Cybernetics* pp. 703–711.

Chan, H. and Ü. Özgüner (1995). Closed-loop control of systems over a communications network with queues. *Int. J. Control* pp. 493–510.

Chizeck, H. J., A. S. Willsky and D. A. Castanon (1986). Discrete–time markovian–jump linear quadratic optimal contrl. *Int. J. Control* pp. 213–231.

Dugard, L. and E. I. Verriest (1997). *Stability and Control of Time-delay Systems*. Academic Press.

Mariton, M. (1990). *Jump Linear Systems in Automatic Control*. New York:Marcel Dekker.

Ray, A. (1988). Integrated communication and control systems : Part I – analysis. *J. Dynamic Systems, Measurement, and Control* pp. 367–373.

Ray, A. (1989). Introduction to networking for integrated control systems. *IEEE Control Systems Mag.* pp. 76–79.

Ray, A. and Y. Halevi (1988). Integrated communication and control systems. part II:design considerations. *J. Dynamic Systems, Measurement, and Control* pp. 374–381.

Copyright © IFAC Distributed Computer Control Systems,
Sydney, Australia, 2000

DESIGNING REAL-TIME SYSTEMS BASED ON MONO-MASTER PROFIBUS-DP NETWORKS [1]

Salvatore Monforte[‡], Mário Alves[†], Francisco Vasques[ξ], Eduardo Tovar[†]

[‡] *IPP-HURRAY Group, on leave from Dept. of Computer Science and
Telecommunications, University of Catania, Italy, E-mail: smonforte@iit.unict.it*
[†] *IPP-HURRAY Group, Polytechnic Institute of Porto, Portugal,
E-mail: {malves@dee, emt@dei}.isep.ipp.pt*
[ξ] *DEMEGI-FEUP, University of Porto, Portugal, E-mail: vasques@fe.up.pt*

Abstract: Profibus networks are widely used as the communication infrastructure for
supporting distributed computer-controlled applications. Most of the times, these
applications impose strict real-time requirements. Profibus-DP has gradually become the
preferred Profibus application profile. It is usually implemented as a mono-master
Profibus network, and is optimised for speed and efficiency. The aim of this paper is to
analyse the real-time behaviour of this class of Profibus networks. Importantly, we
develop a new methodology for evaluating the worst-case message response time in
systems where high-priority and cyclic low-priority Profibus traffic coexist. The proposed
analysis constitutes a powerful tool to guarantee prior to runtime the real-time behaviour
of a distributed computer-controlled system based on a Profibus network, where the real-
time traffic is supported either by high-priority or by cyclic poll Profibus messages.
Copyright © 2000 IFAC

Keywords: Fieldbus Networks, Real-time Communication.

1. INTRODUCTION

A computer-controlled system can be decomposed into
a set of three subsystems: the controlled object; the
computer system; and the human operator (Kopetz,
1997). Collectively, the controlled object and the
human operator can be referred to as the environment
of the computer system.

The role of the computer system is to react to stimuli
from the controlled object or the operator. Basically,
the computer system should be able to receive, via the
instrumentation interface, information about the status
of the controlled object, compute new commands
according to the references provided by the man-
machine interface, and transmit those new commands
to the actuators, also via the instrumentation interface.
A computer-controlled system can have a centralised
architecture, where the field devices (e.g., sensors and
actuators) are connected to the computer system via
point-to-point links. However, there are several
advantages if a field level communication network is
used as a replacement for the point-to-point links. The
main advantage is an economical one. Naturally, the
use of a bus brings other important advantages, such as
easier installation and maintenance, easier detection
and localisation of cable faults, and easier expansion
due to the modular nature of the network. The ability to
support distributed control algorithms is another
advantage achievable by the use of field level
networks.

Typically, a field level network is a broadcast network,
where several network nodes share a common
communication channel. Messages are transmitted
from a source node to a destination node via the shared
communication medium. A major problem occurs
when at least two nodes attempt to send messages via
the shared medium at about the same time. This

[1] This work was partially supported by the European Commission
under the project R-FIELDBUS (IST-1999-11316), by FLAD under
the project SISTER (471/97), and by IDEMEC.

problem is solved by a medium access control (MAC) mechanism.

Most computer-controlled systems are also real-time systems. In general, the issue of guaranteeing real-time requirements is one of checking, prior to run-time, the feasibility of the system's task set; that is, checking if the worst-case execution time of the tasks is smaller than its admissible response time.

In distributed computer-controlled systems, where some of the application tasks are also communicating tasks, it is of paramount importance the evaluation of the messages' response time, since this response time is one of the components of the end-to-end communication latencies.

Therefore, a potential leap towards the use of field level communication networks (fieldbus) in time-critical applications lies in the evaluation of its temporal behaviour.

Profibus (Profibus, 1996) is a well-known fieldbus network that distinguishes between two types of devices - masters and slaves - and supports both mono-master and multi-master systems. A Profibus master is a network device that can send a message on its own initiative, once it gains the right to access the bus. Profibus slaves are devices that may only acknowledge or respond to requests from masters. Generally, they are peripherals such as I/O devices, valves, drives etc.

A widely-used class of Profibus networks is the Profibus-DP (Profibus-DP, 2000) profile. It is generally implemented as a mono-master Profibus network, optimised for speed and efficiency, since it does not implement all the usual communication layers' protocol.

The aim of this paper is to analyse the real-time behaviour of this class of Profibus networks. Importantly, we develop a new methodology for evaluating the worst-case message response time in systems where high-priority and cyclic low-priority Profibus traffic coexist. The proposed analysis constitutes a powerful tool to guarantee prior to runtime the real-time behaviour of a distributed computer-controlled system based on a Profibus network, where the real-time traffic is supported either by high-priority or by cyclic poll Profibus messages.

In (Vasques and Juanole, 1994) the authors provide a real-time analysis of Profibus messages. However, do not consider that Profibus message requests are queued in a FCFS (First-Come-First-Served) queue. Further more, their analysis does not provide any estimation of the worst-case response time of each individual message. In (Tovar and Vasques, 1999a; Tovar and Vasques, 1999b), the authors develop a response time analysis. However, this is intended for multi-master systems and if applied to the mono-master system it would lead to very pessimistic results, since the authors consider always the worst-case token rotation time. Moreover, none of these works, consider the evaluation of response time guarantees for the cyclic poll Profibus messages. This type of messages was analysed in (Li

and Stoeckli, 1993). In this approach, message deadlines are guaranteed since the token cycle time is bounded. The major drawback of this approach is that, in order to evaluate the token cycle time, neither high-priority traffic nor low-priority traffic (other than cyclic traffic) is allowed. This is very restrictive in terms of using Profibus to support real-time distributed computer-controlled applications.

The remainder of this paper is organised as follows. In Section 2 we give a brief description of the main Profibus characteristics. In Section 3 we introduce the network and message models used in the proposed real-time analysis. In Section 4 we propose a methodology to evaluate the Profibus temporal properties, which will be used as a basis for the response time analysis performed in Section 5. In Section 6 a numerical example is given and a realistic scenario of an industrial computer-controlled system is analysed, demonstrating the interest of the proposed analysis. Finally, in Section 7 we draw some conclusions.

2. BASIC CONCEPTS OF PROFIBUS

2.1. Message Cycle

The MAC protocol of Profibus is a simplified version of the timed token protocol (Grow, 1982). The bus access is based on a hybrid method where masters use a token-passing procedure to grant the bus access and a master-slave procedure to communicate with slave stations.

An important Profibus concept is the *Message Cycle*, which comprises the *Action Frame* sent by the *initiator* (always a master) and the associated A*cknowledge* or R*esponse Frame* sent by the *responder*. Profibus allows distinguishing between high-priority, cyclic low-priority (execution of the requests contained in the poll-list) and acyclic low-priority messages. Once the action frame has been transmitted, the initiator waits for the response during a Slot Time (T_{SL}). If a response is not received within T_{SL}, the initiator will try again up to a number of *max_retry_limit* retries.

Profibus provides a service to poll a list of sensors and actuators, by means of a pre-defined sequence of requests. This sequence is called the *Poll List*. The processing of all the *Poll List* entries is said to be a *Poll Cycle*. The *Poll Cycle* duration depends on the length of each message cycle, on the number of message cycles processed at each token arrival and on the token rotation time. Hence, it is obvious that a *Poll Cycle* may last for several token-holding periods. If the *Poll Cycle* is completed within a token holding period, the next *Poll Cycle* may only start at the next receipt of the token. Otherwise, the *Poll List* is processed in segments, without inserting acyclic low-priority message cycles.

2.2. Token Transmission and Reception

The token is passed between masters in ascending order of addresses. To close the logical ring, the master with the highest address passes the token back to the master with the smallest one. In the case of Profibus mono-master networks (Profibus-DP, 2000), the station just passes the token to itself. The advantage of preserving the same token-passing procedure is that they allow for an unambiguous scheduling of different classes of traffic (high-priority and cyclic/acyclic low-priority), preserving all the properties found in multi-master Profibus networks.

2.3. Message Dispatching

At token reception, the period during which the master station is allowed to perform message cycles (*Token Holding Time*) is computed as $T_{TH} = T_{TR} - T_{RR}$, where T_{RR} (Real Rotation Time) is the time between two consecutive token arrivals and T_{TR} (Target Rotation Time) is the expected time for a token cycle.

When a master station receives the token, it processes at least one high-priority message (even if $T_{TH} < 0$). After that, the other pending high-priority message cycles are processed if and while $T_{TH} \geq 0$. It should be pointed out that once a message cycle is started, it is always completed, including any retry (retries), even if meanwhile T_{TH} gets smaller than 0.

The processing of the Poll List is only started after all requested high-priority message cycles have been processed. After each complete Poll Cycle (all entries of the Poll List processed), the requested low-priority message cycles are performed in turn. A new Poll Cycle starts at the next receipt of the token.

3. NETWORK AND MESSAGE MODELS

Requests for message cycles are placed in high-priority, cyclic low-priority or acyclic low-priority outgoing queues. Let $Sh_i^k = (Ch_i^k, Dh_i^k, Th_i^k)$, $Sc_i^k = (Cc_i^k, Dc_i^k, Tc_i^k)$ and $Sa_i^k = (Ca_i^k, Da_i^k, Ta_i^k)$ be high-priority, cyclic and acyclic low-priority message streams in master k, respectively. A message stream is a temporal sequence of requests for message cycles concerning, for instance, the remote reading of a specific process variable.

Ch_i^k, Cc_i^k and Ca_i^k are maximum message cycle duration for a request of message stream Sh_i^k, Sc_i^k and Sa_i^k, respectively. This duration includes the time needed to transmit the request frame and completely receive the related response, and also the time needed to perform the allowed number of message retries. Th_i^k and Tc_i^k are the periodicity of streams Sh_i^k and Sc_i^k requests, respectively. We assume that this periodicity is the minimum interval between two consecutive arrivals of the related requests to the outgoing queue. Dh_i^k and Dc_i^k are the relative deadline of the related message cycle; that is, the maximum admissible time interval between the instant when the message request is placed in the outgoing queue and the instant when

the related response is completely received at the master's incoming queue. Finally, nh^k and nc^k are the number of high-priority and cyclic low-priority message streams, respectively. We also consider that there is just one acyclic low-priority message stream per station.

4. TIMING PROPERTIES OF PROFIBUS

4.1. Token Cycle Properties

In this section, some token cycle properties will be analysed, for the case of Profibus mono-master networks. Namely it will be proved that the real token rotation time T_{RR} is generally smaller than T_{TR} in spite of knowing that once a message cycle is started, it is always completed. More formally, let us introduce the following definitions.

<u>Definition 1</u> – Overrun – We define an overrun as the occurrence of a T_{TH} expiration while a message cycle is being processed.

<u>Definition 2</u> – Overrun Window – We define an overrun window as the time window during which T_{TH} is exceeded due to the completion of a message cycle, added with the subsequent token passing interval.

<u>Definition 3</u> – Late Token – A token is defined as being late if, at its arrival, the real token rotation T_{RR} is greater than the target token rotation time T_{TR}.

A late token arrival implies that at most one high-priority message can be processed by the related master station. It should also be noted that an overrun in a given token arrival, does not usually imply a late token on the next arrival.

Let us denote $A(l)$ as the token arrival instant for the lth token visit. At the time instant $A(l)$, T_{TH}^l is assigned with the value $T_{TR} - T_{RR}^{l-1}$. Therefore, there will be a late token arrival only if $T_{RR}^{l-1} > T_{TR}$. Note that the real token rotation time is measured between token arrivals.

As depicted in Fig. 1, it is clear that the token is neither late in the lth token visit nor in the $(l+1)^{th}$ visit, after the one where an overrun has occurred (as $T_{TR} > T_{RR}^l$). However, in some particular conditions the token can be late.

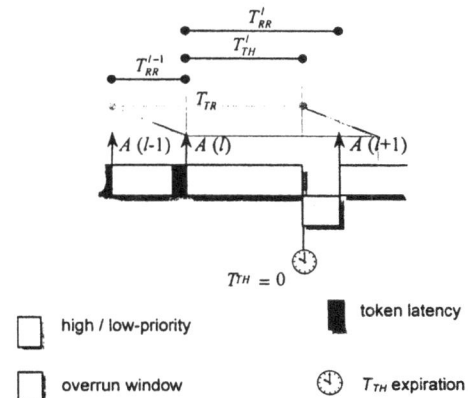

Fig. 1. Overrun and late token.

<u>Theorem 1</u> – In a mono-master Profibus system, if in the lth token visit an overrun occurs, then there will be a late token arrival if and only if the overrun window is greater than the value of T^{l-1}_{RR}.

<u>Proof</u>
Let us assume that in the lth token visit an overrun occurs, and let ω be the overrun window. As shown in Fig. 1, $T^l_{RR} = T^l_{TH} + \omega$. As $T^l_{TH} = T_{TR} - T^{l-1}_{RR}$, then:

$$T^l_{RR} = T_{TR} - T^{l-1}_{RR} + \omega \qquad (1)$$

We must prove that the token will be late at $(l+1)^{th}$ token arrival, that is, $T^l_{RR} > T_{TR}$ if and only if $\omega > T^{l-1}_{RR}$. Since all quantities involved in expression (1) are positive, if $T^l_{RR} > T_{TR}$ then $0 < T^l_{RR} - T_{TR} = -T^{l-1}_{RR} + \omega \rightarrow T^l_{RR} < \omega$, vice-versa if $\omega > T^{l-1}_{RR}$ then obviously follows $T^l_{RR} > T_{TR}$. Hence:

$$T^l_{RR} > T_{TR} \Leftrightarrow \omega > T^{l-1}_{RR}$$

That is, a late token arrives at the $(l+1)^{th}$ token visit if and only if the length of the overrun window is greater than the value of T_{RR} computed at the beginning of the lth cycle, that is, is greater that T^{l-1}_{RR}. ❑

<u>Corollary 1</u> – A low-priority overrun induces a late token if and only if in the previous token visit no low-priority and at most one high-priority message has been processed.

<u>Proof</u>
For a low-priority overrun the length of the longest overrun window is $\omega = Cl_{max} + \tau$, where $Cl_{max} > Ch_{max}$.

If in the previous token visit, no low-priority messages and at most one high-priority message were processed, at the token arrival $T_{RR} \leq Ch_{max} + \tau$, that is smaller than ω and thus a late token will be induced on next visit. ❑

4.2. Basic Response Time Analysis

In (Liu and Layland, 1973), the author introduced the concept of critical instant as being the time instant at which a request for a given task has the longest response time, that is, the longest time interval till the end of the response for that request. Moreover, a critical instant for any task occurs whenever the task is requested *simultaneously* with requests for all higher priority tasks (Liu and Layland, 1973).

In a Profibus network, we must consider that, due to the FCFS behaviour of the outgoing queues, a given message request can be delayed by requests from all the other message streams (contrarily to the task scheduling case where it would be delayed only by high priority message stream requests). Therefore, a critical instant will occur when, for a given priority (i.e. high- or acyclic low-priority) every message stream simultaneously issue a message request.

Moreover, due to the non pre-emptive context of messages processing, high-priority request may suffer some additional delay before starting being processed. Let us introduce the following definitions:

<u>Definition 4</u> – Profibus Critical Instant – Considering that requests for all high-priority, cyclic and acyclic low-priority message streams are simultaneously placed on the respective outgoing queues. We define a Profibus critical instant as the time instant at which a request for a given message stream has the longest response time.

<u>Definition 5</u> – Initial Blocking – We define the initial blocking as the delay that the first request made at the critical instant may suffer until starting to be processed.

<u>Definition 6</u> – Critical Load – We define the critical load for a given priority class, as the time interval between a critical instant and the time instant when the last request (made at the critical instant) for that priority class has been completely processed.

As far as the evaluation of the worst-case response time is concerned, two factors must be taken into account: the initial blocking and the high-priority critical load. The worst-case response time for both high-priority and low-priority messages stream made at the critical instant results from the simultaneous occurrence of:
1. the longest initial blocking, that is, the first high-priority request suffers the longest possible delay before being processed;
2. the longest high-priority critical load, that is, it takes the maximum number of token visits to process all high-priority requests.

The following two theorems prove that the simultaneous occurrence of both conditions leads to the worst-case response time for the last message request to be processed.

<u>Theorem 2</u> – The longest initial blocking occurs when all requests are issued simultaneously with the occurrence of a low-priority overrun.

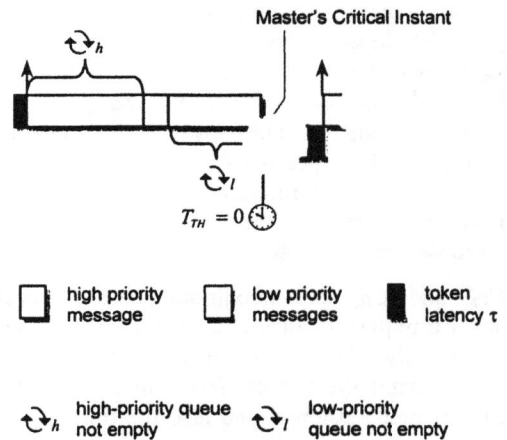

Fig. 2. Master's Critical Instant.

<u>Proof of Theorem 2</u>
Considering that the initial blocking depends on the position of the critical instant itself, the theorem is proved if a shifting of its position leads either to a deadline violation or to a smaller blocking.

Referring to Fig. 2, if the position of the critical instant is anticipated (shifted to the left), this leads to a deadline violation, since it means that requests were issued while the queue was not empty (either for low-priority or high-priority).

Finally, if the critical instant is postponed (shifted to the right), this leads to a smaller blocking. ❑

Corollary 2 – The longest initial blocking is $B = Cl_{max} + \tau$.

Proof
From Fig. 2, it is clear that the longest low-priority overrun is $Cl_{max} + \tau$. ❑

The length of the high-priority critical load interval depends not only on the occurrence of the longest initial blocking, but also on the traffic processed on the previous token cycle.

Theorem 3 – A Profibus critical instant occurs when the longest initial blocking is preceded by a token visit where no low-priority and at most one high-priority message is processed.

Proof
As depicted in Fig. 3, a late token arriving after a Profibus critical instant leads to a longer high-priority critical load, since the second high-priority request to be processed will suffer a delay of τ. Taking into account that the maximum initial blocking is equal to the longest low-priority overrun window (Corollary 2); and in view of Corollary 1, after the critical instant a late token arrives if no low-priority message and at most one high-priority message is processed in the previous token visit. ❑

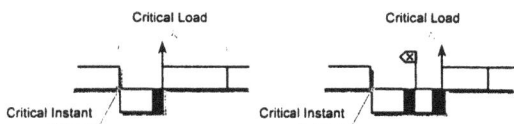

Fig. 3. Late token and high-priority critical load.

Corollary 3 – The maximum high-priority load interval leads to the worst-case response time for both cyclic and acyclic low-priority message streams.

Proof
Considering that low-priority message cycles are performed in turn only after all high-priority message have been processed, its is clear that the maximum high-priority load interval leads to the maximum span of time between the critical instant and the time instant at which the last low-priority message (either cyclic or acyclic) is processed; and thus to the worst-case response time evaluation. ❑

4.3. Processing of High Priority Messages

Since all the possible requests are issued at the critical instant and due to the non pre-emptive context of Profibus, a master may need several token visits to process all high-priority messages, before processing any low-priority request. Thus, there will be a well-defined pattern when processing all those requests (Fig. 4).

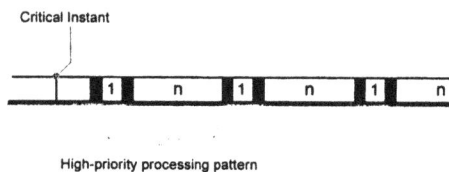

Fig. 4. The 1-n processing pattern.

This processing pattern is characterised by a late token arrival, where just *one* high-priority message is processed, followed by an early token arrival, where n high-priority messages are processed.

Theorem 4 – The occurrence of a Profibus critical instant induces a 1-n processing pattern for high-priority messages.

Proof
From the analysis, it follows that after the occurrence of a critical instant, a late token arrives. Thus, as depicted in Fig. 5, only one high-priority message can be processed (❶) and on next token visit T_{TH} is assigned with a maximum values equal to (T_{TR} - Ch_{max} - τ). Therefore, if the number of high-priority requests issued at the critical instant is greater than the number of high-priority messages which can be processed

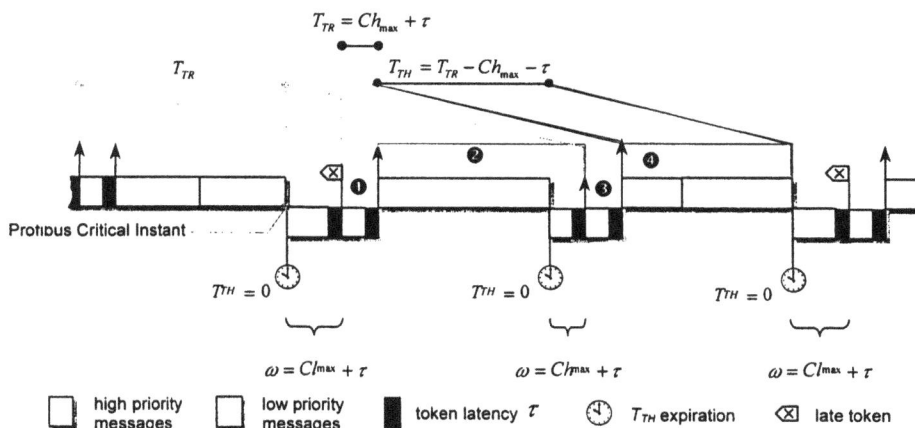

Fig. 5. Evaluation of the 1-n high-priority processing pattern.

during T_{TH}, then a high-priority overrun occurs, and, in the worst case, the length of the overrun window is $(Ch_{max} + \tau)$. Consequently, the token real rotation (T_{RR}) equals the target token rotation time (T_{TR}) (❷), and on next token visit only one message cycle will be carried out (❸). The 1-n is clearly defined, where $(n -1)$ messages are processed before the expiration of T_{TH} and the last of the n messages is processed in overrun. The maximum time spent processing these n messages is equal to $(T_{TR} - Ch_{max} - \tau)$ for processing $(n -1)$ messages, plus $(Ch_{max} + \tau)$ to process the message in overrun, that is, is equal to T_{TR}. It is also clear that if:

- the smallest token cycle always comprises one high-priority message cycle; and
- in a token cycle where at least one low-priority message is processed the token is never released before the expiration of the token holding time timer (T_{TH});

then, in the last token cycle where there are still high-priority request pending, the token holding timer is assigned with $T_{TH} = T_{TR} - Ch_{max} - \tau$ (❹), and this occurs after a token cycle which comprise no low-priority and exactly one high-priority message cycle (❶, ❸). Moreover, if an overrun of T_{TH} occurs then in the next token cycle no low-priority and at most one high-priority message can be processed (❺). ❑

5. WORST-CASE RESPONSE TIME

5.1. High-priority Message Streams

The worst-case response time for high-priority message can be computed taking into account the following 4 components:
1. the initial blocking;
2. the time spent processing requests in token visits where n high-priority messages can be processed, plus the time spent to pass the token, that is, $\lfloor nh/n \rfloor \times T_{TR}$;
3. the time spent processing requests in token visits where just 1 high-priority message can be processed, plus the time spent to pass the token, that is, $\lfloor nh/n \rfloor \times (Ch_{max} + \tau)$;
4. a component Ψ_h, which is related to the finishing of the 1-n processing pattern. We consider that at the end of the last complete cycle of n messages, there are three possible cases:
   a) there are no more pending requests, and thus the computation of the response time ends before releasing the token in the previous token cycle;
   b) there is just one pending request, and thus the time needed to process the pending requests is exactly Ch_{max};
   c) there are more than one pending request, and thus one more token cycle is needed to process the pending requests.

Hence, the worst-case response time for Profibus high-priority messages is:

$$R_h = B + \lfloor nh/(n+1) \rfloor \cdot (T_{TR} + Ch_{max} + \tau) + \psi_h \qquad (2)$$

where:

$$\psi_h = \begin{cases} -\tau, \text{ if } nh = \lfloor nh/(n+1) \rfloor \cdot (n+1) \\ Ch_{max}, \text{ if } nh = \lfloor nh/(n+1) \rfloor \cdot (n+1)+1 \\ (nh - \lfloor nh/(n+1) \rfloor \cdot (n+1)) \cdot Ch_{max} + \tau, otherwise \end{cases}$$

and

$$n = \lfloor (T_{TR} - Ch_{max} - \tau)/Ch_{max} \rfloor + 1 = \lfloor (T_{TR} - \tau)/Ch_{max} \rfloor$$

5.2. Cyclic Low-Priority Message Streams

The processing of low-priority requests issued at the critical instant, is alternated by periods where new (those requested during the processing of low-priority messages) high-priority message requests are carried out (Fig. 6).

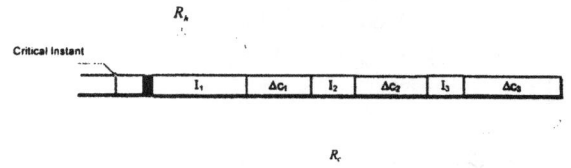

Fig. 6. Interleaving of interference intervals and cyclic processing intervals.

<u>Definition 7</u> – Interference Interval – We define an interference interval I_i as the ith time window during which only high-priority messages are processed.

<u>Definition 8</u> – Cyclic Processing Interval – We define a cyclic processing interval ΔC_i as the ith time window during which low-priority traffic is processed.

It is clear that considering both cyclic processing intervals and interference intervals it is possible to evaluate the worst-case response time for cyclic-low priority message streams.

The length of the ith cyclic processing interval can be computed as the difference between the value of $T_{TR} = T_{TR} - Ch_{max} - \tau$ and the maximum amount of time used to process the remaining high-priority messages, that is:

$$\Delta c_i = (T_{TR} - Ch_{max} - \tau) - \\ - \max(0, (n_i^h - \lfloor n_i^j/(n+1) \rfloor \cdot (n+1))-1) \cdot Ch_{max} + Cl_{max} + \tau$$

where n_i^h is the number of high-priority messages processed in I_i.

It should be noticed that every high-priority request which arrives within ΔC_i will be pending on next token cycle (start of I_i.). Moreover, there is a mutual dependence between the evaluation of the number n_i^h of high-priority messages processed in the ith interference interval and the length of this interval itself:

$$\sum_{j=1}^{nh} \left\lfloor \left(I_i + \sum_{m=1}^{i-1} (I_m + \Delta c_m) + B \right) \middle/ Th_j \right\rfloor \qquad (3)$$

The interference imposed by the processing of n_i^h high-priority messages is:

$$I_i = \left\lfloor \frac{n_i^h}{n+1} \right\rfloor \cdot (T_{TR} + Ch_{max} + \tau) + Ch_{max} + \tau +$$

$$+ \max\left(0, \left(nh_i - \left\lfloor \frac{n_i^h}{n+1} \right\rfloor \cdot (n+1) \right) - 1 \right) \cdot Ch_{max}$$

Finally the worst-case response time for cyclic low-priority message streams can be expressed as follows:

$$R_c = B + \sum_{i=1}^{m-1} (\Delta c_i + I_i) + I_m + \left(nc - \sum_{i=1}^{m-1} n_i^c \right) \cdot Cl_{max} \quad (4)$$

where $n_i^c = \lfloor (\Delta C_i - \tau) / Cl_{max} \rfloor$ is the number of cyclic low-priority messages in ΔC_i and $m = \min\{n \in \aleph : \sum_{i=1,...,n} n_i^c \geq nc\}$.

6. NUMERICAL EXAMPLE

In this section a realistic scenario for an industrial computer-controlled system is utilised to demonstrate the interest of the proposed analysis.

Consider a mono-master Profibus network in an industrial environment, where a decentralised computer-controlled system integrates video message streams, computer-generated audio messages and control-related message streams. The supported application controls an assembly production line where parts must be assembled and checked. Both completeness tests (i.e. check if all parts are complete and in position) and dimension tests (i.e. verifying if parts are within the prescribed tolerance for diameters, distances, radii, angles etc.) are required for the assembly process. An intelligent stand-alone vision systems is used to directly evaluate images according to the stored testing program. Thus, only the data resulting from the video inspection operation will be sent through the Profibus interface.

The control-related message streams, which interconnect sensors and actuators to controllers, are mapped on the high-priority message streams. Table 1, summarises the characterisation of the high-priority message streams considered in this example, where set x represents a set of message streams with the same periodicity.

Table 1 Summary of high-priority message streams

	Set 1	Set 2	Set 3	Set 4
nh=20	3	5	7	5
Th_i	20 ms	25 ms	50 ms	60 ms

In order to meet realistic requirements, a value for the Slot Time (T_{SL}) and for the target token rotation time (T_{TR}) has been fixed to 100µs and 8ms, respectively. Moreover, a 1.5Mbps data transfer rate and no retry ($\rho = 0$) are considered in the example. Therefore, the bit period is equal to 0.667µs and the computation for

the token-passing latency, yields: $\tau = 3 \times (T_{TF} + T_{SL}) = 3 \times (22 + 100) = 0.366$µs, where T_{TF} is the token frame length and 3 is the maximum number of retries predefined for the case of the token frame.

Concerning message streams to support the video inspection capabilities, we assume that the maximum data size for the result of an image evaluation is 246 bytes (equal to the maximum length for cyclic low-priority messages) and the image shutter rate is 20 picture/s. These requirements leads to polling each camera device every 50ms.

Concerning the multimedia services, which are also mapped over cyclic low-priority message streams, we assume 128Kbps as the bandwidth requirement. Hence, for a multimedia video stream application with 352x288 frame format at 10 frames/s, each video device must be polled every 15ms.

Table 2 summarises the cyclic low-priority message streams characterisation considering the utilisation of both 5 intelligent Profibus cameras for part inspection and 2 surveillance video cameras.

Table 2 Summary of cyclic message streams

	Set 1	Set 2
nc=7	2	5
Tc_i	15 ms	50 ms

Table 3 Message cycles length

	bytes	length
Ch_{max}	20 bytes	0.433ms
Cl_{max}	246 bytes	1.569ms

Considering the length for cyclic low-priority message cycle presented in Table 3, the maximum initial blocking is: $B = Cl_{max} + \tau = 1.935$ms. Thus, the number of high-priority message processed in a 1-n processing pattern is: $n + 1 = \lfloor (T_{TR} + Ch_{max} - \tau) / Ch_{max} \rfloor = \lfloor 18.617 \rfloor = 18$. Therefore, the worst-case response time for high-priority message streams is:

$$R_h = B + \left\lfloor \frac{nh}{n+1} \right\rfloor \cdot (T_{TR} + Ch_{max} + \tau) + \psi_h =$$

$$= 1.935\ ms + \left\lfloor \frac{20}{18} \right\rfloor \times 8.799\ ms + 1.232\ ms = 11.967\ ms$$

Thus, the high-priority message stream set is schedulable as $R_h < 15$ms $= \min\{Th_i\}$.

Let us now compute the worst-case response time for cyclic low-priority message streams. The first interference interval is the one imposed by processing the nh high-priority requests made at the critical instant:

$$I_1 = \left\lfloor \frac{nh}{n+1} \right\rfloor \cdot (T_{TR} + Ch_{max} + \tau) + Ch_{max} + \tau +$$

$$+ \max\left(0, \left(nh - \left\lfloor \frac{nh}{n+1} \right\rfloor \cdot (n+1) \right) - 1 \right) \cdot Ch_{max} =$$

$$= 10.032\ ms$$

The length of the first cyclic processing interval is:

$$\Delta c_1 = \left(T_{TR} - Ch_{max} - \tau\right) - $$
$$- \max\left(0, \left(nh - \left\lfloor \frac{nh}{n+1}\right\rfloor \cdot (n+1)\right) - 1\right) \cdot Ch_{max} + $$
$$+ Cl_{max} + \tau = 8.703 \ ms$$

The number of cyclic low-priority requests that can be accomplished in ΔC_i is then:

$$n_1^c = \left\lfloor \frac{\Delta c_1 - \tau}{Cl_{max}}\right\rfloor = \left\lfloor \frac{8.703 \ ms - 0.366 \ ms}{1.569 \ ms}\right\rfloor = \lfloor 5.312 \rfloor = 5$$

Hence, there are still 2 cyclic pending requests. It should be reminded that there is a mutual dependence between the evaluation of the number n_i^h of high-priority messages processed in the ith interference interval and the length of the interval itself. Hence, in order to compute n_2^h we need to know the value of I_2 and vice versa. From equation (3), it follows that:

$$n_2^h = nh + \sum_{j=1}^{nh}\left\lfloor \frac{I_2 + (I_1 + \Delta c_1) + B}{Th_j}\right\rfloor - nh = $$
$$= \sum_{j=1}^{nh}\left\lfloor \frac{I_2 + 20.67 \ ms}{Th_j}\right\rfloor$$

Hence:

$$I_2 = \left\lfloor \frac{n_2^h}{18}\right\rfloor \times 8.799 \ ms + 0.799 \ ms + $$
$$+ \max\left(0, \left(n_2^h - \left\lfloor \frac{n_2^h}{18}\right\rfloor \times 18\right) - 1\right) \times 0.433 \ ms$$

The easiest way to solve such dependence is to form a recurrence relationship:

$$n_2^{h[m+1]} = \sum_{j=1}^{nh}\left\lfloor \frac{\left\lfloor \frac{n_2^{h[m]}}{18}\right\rfloor \times 8.79 + 0.79 + \Theta \times 0.43 + 20.67}{Th_j}\right\rfloor,$$
$$\Theta = \max\left(0, \left(n_2^{h[m]} - \left\lfloor \frac{n_2^{h[m]}}{18}\right\rfloor \times 18\right) - 1\right)$$

It follows that $n_2^{h[0]}=0$, $n_2^{h[1]}=3$ and $n_2^{h[2]}=3$. The iteration stops since $n_2^{h[1]}=n_2^{h[2]}$ and thus, $I_2 = 0.799 + 2 \times 0.433 = 1.666ms$. The length of the second cyclic processing interval is $\Delta C_2 = (T_{TR} - Ch_{max} - \tau) + 2 \times Ch_{max} + Cl_{max} + \tau = 8.269ms$, which allows the processing of a number of cyclic low-priority messages is given by: $n_2^c = \lfloor (\Delta C_2 - \tau) / Cl_{max}\rfloor = \lfloor(8.269 - 0.366) / 1.569\rfloor = \lfloor 5.036 \rfloor = 5$. Therefore, considering that $nc^c < n_1^c + n_2^c$ the computation of worst-case response time for cyclic low-priority message streams is over and yields:

$$R_c = B + \Delta c_1 + I_1 + I_2 + \left(nc - n_1^c\right) \cdot Cl_{max} = $$
$$= 20.67 \ ms + 1.666 \ ms + 3.139 \ ms = 25.475 \ ms$$

7. CONCLUSIONS

Profibus networks are often used as the communication infrastructure for supporting distributed computer-controlled applications. Generally, these applications impose strict real-time requirements. A potential leap towards the use of Profibus in time-critical applications lies in the evaluation of its temporal behaviour.

A widely-used class of Profibus networks is the Profibus-DP profile, usually implemented as a mono-master Profibus network. In this paper we analysed the real-time behaviour of this class of Profibus networks. Importantly, we developed a new methodology for evaluating the worst-case message response time in systems where high-priority and cyclic low-priority Profibus traffic coexist. The results presented in this paper constitute an important advance concerning the state-of-the-art in the sense that previous relevant works neither could be applied to the mono-master case nor considered the response time analysis for systems supporting both high priority and cyclic poll Profibus messages.

Work is now being carried out in order to extend the approach followed in this paper to multi-master Profibus networks.

8. REFERENCES

Grow, R. (1982). A Timed Token Protocol for Local Area Networks. In: *Proceedings of Electro'82*, Token Access Protocols, paper 17/3.

Kopetz, H. (1997). Real-Time Systems: Design Principles for Distributed Embedded Applications. Kluwer Academic Publishers.

Li, M. and L. Stoeckli (1993). The Time Characteristics of Cyclic Service in Profibus. In: *Proceedings of the EURISCON'94*, Vol. 3, pp. 1781-1786.

Liu, C. and J. Layland (1973). Scheduling Algorithms for Multiprogramming in a Hard Real-time Environment. In: *Journal of the Association for Computer Machinery*, (20)1, pp. 46-61.

Profibus (1996). General Purpose Field Communication System, Volume 2, EN 50170.

Profibus-DP (2000). Profibus Technical Overview. http://www.profibus.com.

Tovar, E. and F. Vasques (1999a). Cycle Time Properties of the PROFIBUS Timed Token Protocol. In: *Computer Communications*, Elsevier Science, 22(13), pp. 1206-1216.

Tovar, E. and F. Vasques (1999a). Real-Time Fieldbus Communications Using PROFIBUS Networks. In: *IEEE Transactions on Industrial Electronics*, 46(6), pp. 1241-1251.

Vasques, F. and G. Juanole (1994). Pre-run-time Schedulability Analysis in Fieldbus Networks. In Proceedings of IECON'94, Bologna, Italy, pp. 1200-1204.

Copyright © IFAC Distributed Computer Control Systems,
Sydney, Australia, 2000

A MONITOR TOOL FOR NETWORKED FACTORIES

Klaus Kabitzsch [1], Gerolf Kotte [2]

[1] *Dresden University of Technology, Faculty of Computer Science*
[2] *Systema Cell Automation, Software Division, Dresden (Germany)*

Abstract: The following tool helps the troubleshooters in networked factories. It records time-dependent operations and makes these visibly. It offers methods to the abstraction and condensation and helps, to find anomalies. *Copyright © 2000 IFAC*

Keywords: Communication Protocols, Computer-aided diagnosis, Discrete-event systems, Fieldbus, Logic analysers, Performance monitoring, Sequence estimation

1. INTRODUCTION

If the complexity increases in factories because of the use of computers, networks and PLC's, many disturbances and real time problems appear at the time of commissioning. Errors appear only rarely and nobody can observe the faulty operations with bare eye, because they are too fast. The suppliers of the systems record therefore all processes in the computers and networks over longer time. During the recording chronological data series emerge with time tag. These recordings are also called history or trace (Ludwig, *et al.*, 1996). Such history file can emerge in a factory every day with a size up to 200 Mbyte. For this reason most operators have, however, no insight in their installations (Kabitzsch and Vasyutynskyy, 1999).

Table 1 Different sources for a recording

Place of the measurement	Tools for the measurement
Networks , fieldbuses , serial communication	Protocol analyzer
Electronics , hardware	Logic analyzer
Operating systems	System monitor
Data bases	Transaction - protocol

User - programs	Debugger , trace – monitor
Real time controls, PLCs, teleservice at machines	Service - monitor
Sensors, actuators	Data logger
Simulators (offline, online, hardware in the loop)	Simulation - protocol

Extrakt is a software - tool used for the detailed analysis of chronological recordings. These results were measured previously with different measuring - tools. If Extrakt receives the histories as data base- or Excel- files, it can process these immediately. Other formats are converted with the tool TraceKonverter into a data base. Extrakt can process histories of arbitrary length. The only limit are the resources of the available PC.

2. EXAMPLE OF RECORDINGS AT A MACHINE

Following simple example describes the application of Extrakt. A transport system in a factory is controlled by four computers. These are joined by a network and exchange variable - updates (Figure 2).

The progress of the transport operations can be measured as traffic of transactions at the network (Figure 5). The history recorded by the protocol monitor is converted into a data base and presented by Extrakt as a table. Figure 5 shows the following columns of the table:

"Zaehler" — number of the measured telegram
"Zeit" , "ms" — time stamps (date, time of day, ms)
"Service" — communication service in the network (Ackd, UnAckd, ACK)
"Absender" — names of the computer nodes,
"Empfaenger" — which participate in the communication
"NV" — names of the variable, their update will transfer (network variable)
"DataNum", — values (contents) of the respective
"DataAsc" — updates

The tables - view (Figure 5) shows the events for only a small section of the history (only few data records). The user can divide the table for this reason horizontally and vertically. The new windows have variable size, because the limit - beams are relocatable with the help of the mouse. The two horizontal windows indicate different periods of the recording. In this way two operations can be compared, which were measured to different times. The vertical division is necessary, if very many table columns exist, which find no place on the screen. The screen contains still only a small section of the recording. Extrakt offers of course a help to the lookup, which quickly finds a desired data record. For practical measurements files range from 10 to 200 Mbyte. Then all details of the transport operations are shown, but an overview over larger periods is impossible. The table does not provide details about the actual time ratios.

3. TIME ANALYSIS

The time analysis gives a better view of the real time relations in the history. All processes are presented in real time, spread over a linear time axis. First Extrakt analyzes in the background the diversity of contents and values in the fields of a table column. For example Extrakt has observed in the fields of the column "NV" the variables nv_ref_pkt_in , nvi_ref_pkt , nv_smotor_data , nv_pos_update ... (Figure 5) etc. In the same sense Extrakt analyzes also the values - volumes of the other columns of the table. Figure 3 is a copy of the time analysis window. The user chooses a table column and Extrakt shows the list of analyzed values - volume (detail 4). The horizontal axis (detail 2) is the linear time, which was calculated from the time stamps. The sample represents over the time axis by a stroke "l", to which time a corresponding content was recorded in the history. In the example (column "NV") the lines (variables) show, when an update would transfer for the corresponding variable in the network. Each

transaction is presented only as small symbol "l". Therefore the user can overlook recordings up to 1000 transactions on a look. One can get an impression of the actual time proportions. Extrakt can convert the values - volumes of the other columns into comparable representations. The viewer gets thereby respectively other sighting on the operations. The time analysis window allows:

- The rolling of the visible indication - range (window) moves the image to the left or to the right
- The zooming of the time scale is possible
- The sequence of the values (detail 4) can be exchanged
- New lines can be loaded in the window
- A double click on the symbols "l" shows the complete information of the table to the data
- Groups of events are presented colorfully
- A time - cursor can change immediately to the same time in the tables - view (Figure 5)

4. SUMMARIZING THE HISTORY

The time analysis creates a compact representation of the history. But if files must be analyzed with many thousand data records, the survey goes still lost. As soon as the number of events becomes too large, each event no longer can be presented as separate symbol "l". Adjacent events then can be replaced from a common groups - symbol "!". In this case the resolution in the detail of the representation is lost. This resolution can be restored by zooming, but then it will lose the overview over the entire file. To solve this problem, Extrakt offers many abstraction - methods. The user has the choice, which details him are important in the desired view. He chooses thereby simultaneously, on which details he dispenses, to get a better total - overview.

4.1 Filters

If Extrakt indicates only the data records, which correspond to the filter, it can emphasize interesting details from the data base. The presentation is condensed leaving out unimportant details. Extrakt can define arbitrarily many filters. If the user is interested in the communication between the nodes "Sensorik" (sensors) and "Ablaufsteuerung" (control) of our transport system only, the corresponding transactions should be indicated only:

1) from the transmitter (Absender) "Sensorik" to the receiver (Empfaenger) "Ablaufsteuerung"
2) from the transmitter (Absender) "Ablaufsteuerung" to the receiver (Empfaenger) "Sensorik"

Extrakt offers a form for the comfortable definition (specification) of the filters. The condensed results are presented in the tables - view and in the time analysis windows.

4.2 Events describe and search

Events appear in the system, if according to the opinion of the troubleshooter "something important happens" (Sreenivas and Krogh, 1991). Because the troubleshooter wants to work during his future analyses with these events, he gives them names. With this vivid names he can then address and use the events in Extrakt. The start of a transport process is an important event. The user gives it the name "Einlochen beginnt" (transport begins). This event is activated by the reference point sensor (Figure 2) and is tied with all data records, which have following features:

Transmitters (Absender) = "Sensorik"   AND
Receivers (Empfaenger) = "Ablaufsteuerung" AND
Variable (NV)   = "nv_ref_pkt_in"

If the bead falls in the hole, this is also an important event in the transport system. These events get future the name "Kugel im Loch" (bead in the hole). This name is tied with all the data records, which have the following features:

Transmitters (Absender) = "Sensorik"   AND
Receivers (Empfaenger) = "Ablaufsteuerung" AND
Variable (NV)   = "nv_proz_erg_in"

Extrakt offers a window (template) for the comfortable definition (specification) of the events. The results are presented in tables - view (Figure 5). Extrakt adds new columns to the table and marks them with the defined names of the events. If a data record corresponds to a specified (described) event, then „E_" appears in the corresponding column.

4.3 Sequences describe and search

Extrakt should search not only individual events but also complex operations, processes, performance - patterns in the transport system, which are important in the opinion of the troubleshooter. These patterns should get discrete names, with which the user works in the future. Each complete, correct transport - process of the bead from beginning to end is an important behavior pattern in the transport system. This pattern gets the name "Erfolgreiches Einlochen" (successful transport) and is marked by a sequence of three events as shown in figure 4. First requirement for the formal description of the pattern is the description of all previously involved events in Extrakt. Then the formal names of these events are usable for new descriptions. For this purpose Extrakt offers a form with a list of all previously defined events. Events are added gradually with the mouse. The new sequence is assembled in this way. If this definition becomes active, the display looks as shown in figure 5. Extrakt adds to the table further

columns and gives them the names of the new sequence - patterns. In the new columns a „R" appears, if a data record was found, which belongs to the new patterns. A factory has many complex performance - patterns, in which several sequences operate successively. The user must describe therefore "higher level" patterns, which are put together from a defined sequence of "lower level" patterns (sequences). In this sense the troubleshooter can use the form for the formal description of sequences recursively. Extrakt can define an arbitrary number of hierarchy - levels and search very complex patterns in the data base. Each new pattern gets its own formal name. An example may be a hierarchy from the control of a high shelf warehouse. In the example the storage of workpieces is observed. The lowest level shows the traffic of the transactions in the fieldbus. The next level shows, which services the PLCs call successively. The most abstract levels contain the service calls in the overstocked workstation (manufacturing executive system).

4.4 Representation of the abstraction in the time analysis window

The time analysis window (Figure 3) also allows the representation of all abstract elements (events and sequences). If abstract elements were defined formally, then their new names are mentioned in the list (detail 4). In the pertinent lines over the time – axis, Extrakt presents all corresponding events, which are found in the history. If these elements correspond to the formal description, they are presented by the symbol " I ". If the resolution is not sufficient and many " I " - symbols lie near other, group icons are used. Sequences become marked with the symbol I:::I::::I . The user can choose now between total overview and single detail. He can determine which levels of the formal hierarchy elements to appear. He can change with few selections between the levels of the abstraction. Mostly he begins with a general survey and makes finer resolutions as he progresses.

5. INTERPRETATION WITH INSIGHTS AND STATISTICS

The user can create insights and statistics with few selections. In our example this corresponds to the burden (traffic) in the network. One can determine the correlations between the contents of two columns. They result from laws of nature or from the tasks - specification of the development engineer or programmer. The correlation analysis is used to detect dependence in metric signals and linear systems. The value - to - value - statistics is applicable, if the data base contains nominal values.

If the data base contains two columns (X , Y) with metric values, Extrakt can enter all values - pairs (x , y) as points in a diagram y=f(x). Thereby point clouds emerge and show functional dependence. If the troubleshooter uses the possibilities of the abstraction (events, sequences), he can also get insights and statistics in reference to these formal elements. In this case Extrakt calculates the frequencies of events, sequences etc. Extrakt can also calculate the term and the time interval of sequences and events. Mostly the diagnosis begins first, after disturbances (effects) were observed at the test or while the factory is running. The search for the belonging errors (causes) begins with monitoring. If time-dependent and transient effects were observed, the interpretation requires three steps, which capture the knowledge about the disturbances:

(1) Specification of the error effects: First all disturbances (effects) are described formally. All methods are recursive as described above. Thus hierarchies of events and sequences (scenarios) are defined.

(2) Search for similar error effects: Extrakt searches the data base with the help of these specifications. Thereby all similar scenarios are found, which have operated in the factory. The tool finds even similar scenarios, which were not noticed by the users.

(3) Segmentation of the periods: The entire period of the recording is divided now without gaps in segments. An attribute is associated to each segment, which was selected from 4 variants.

(4) Search for mutualities and differences: To track the anomalies, pattern recognition procedure is utilized for all segments. Signal analysis, static and dynamic systems analysis (Köppen-Seliger, et al.,

1999; Kotte and Kabitzsch, 1999) compare correct and disturbed segments and reveal dependence in the interior of the system.

REFERENCES

Sreenivas, R.S. and B.H. Krogh (1991). On condition/event systems with discrete state realizations. In: *Discrete event dynamic systems: theory and applications* 1 (2), 1991, pp. 209-236

Ludwig, T., R. Wismüller and A. Bode (1996). OMIS-On-line monitoring interface specification. *Report N342/05/96A Sonderforschungsbereich 342*, TU München

Köppen-Seliger, B., E. Alcorta Garcia and P.M. Frank (1999). Fault detection: Different strategies for modeling applied to the three tank benchmark. In: *Proc. ECC'99 European Control Conference, Karlsruhe* , Part CA-5

Kabitzsch, K. and V. Vasyutynskyy (1999). Tele-Diagnosis at Networked Automation Systems. In: *FeT'99 Fieldbus Technology Conference,* Magdeburg, Sept. 1999, Proceedings pp. 209-214, Springer Verlag Wien

Kotte, G. and K. Kabitzsch (1999). Monitoring in Semiconductor Manufacturing. In: *ECC'99 European Control Conference*, August 1999, Karlsruhe, Proceedings Part CM-1

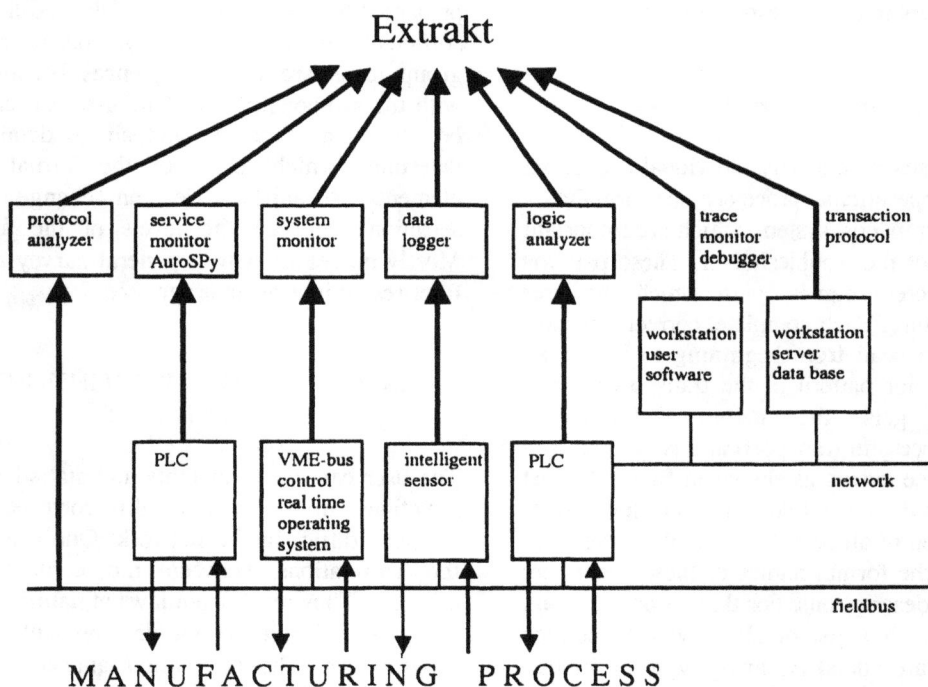

Fig. 1. Tools and monitor operations in a factory

| node **„Ablaufsteuerung"** control of operation | node **„Schrittmotor"** stepping motor | node **„Sensorik"** sensors | node **„Diagnose"** diagnosis |

fieldbus / network

sends the variables:

stepping motor forward (A)
(nv_smotor_data)

timer expired (detail E)
(nv_delay_stdin)
(nvi_tout_erg)

sends the variables:

turn angle position (A)
(nv_pos_update)

sends the variables:

reference point sensor (detail B)
(nv_ref_pkt_in)
(nvi_ref_pkt)

speed sensor (detail C)
(nvi_gsensor)
(nv_delay_in)

result sensor (detail D)
(nv_proz_erg_in)
(nvi_erg_sensor)

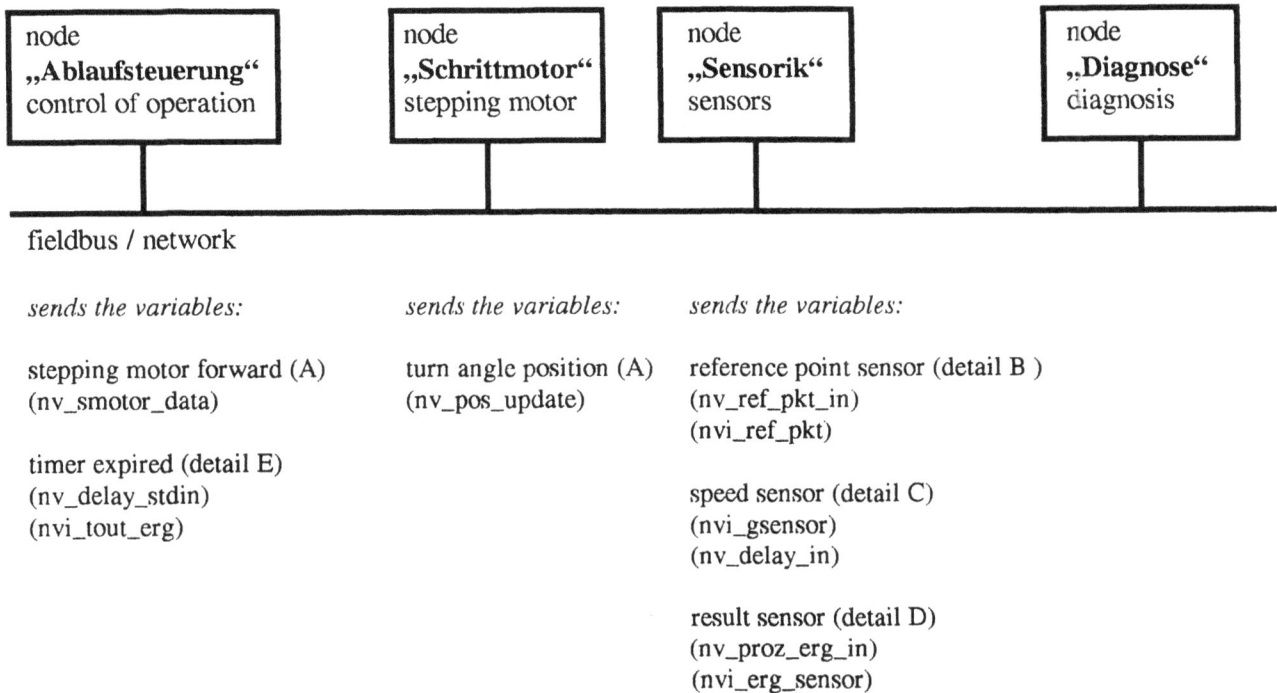

Fig. 2. Principle of the networked control (details A, B, C, D, E see figure 3)

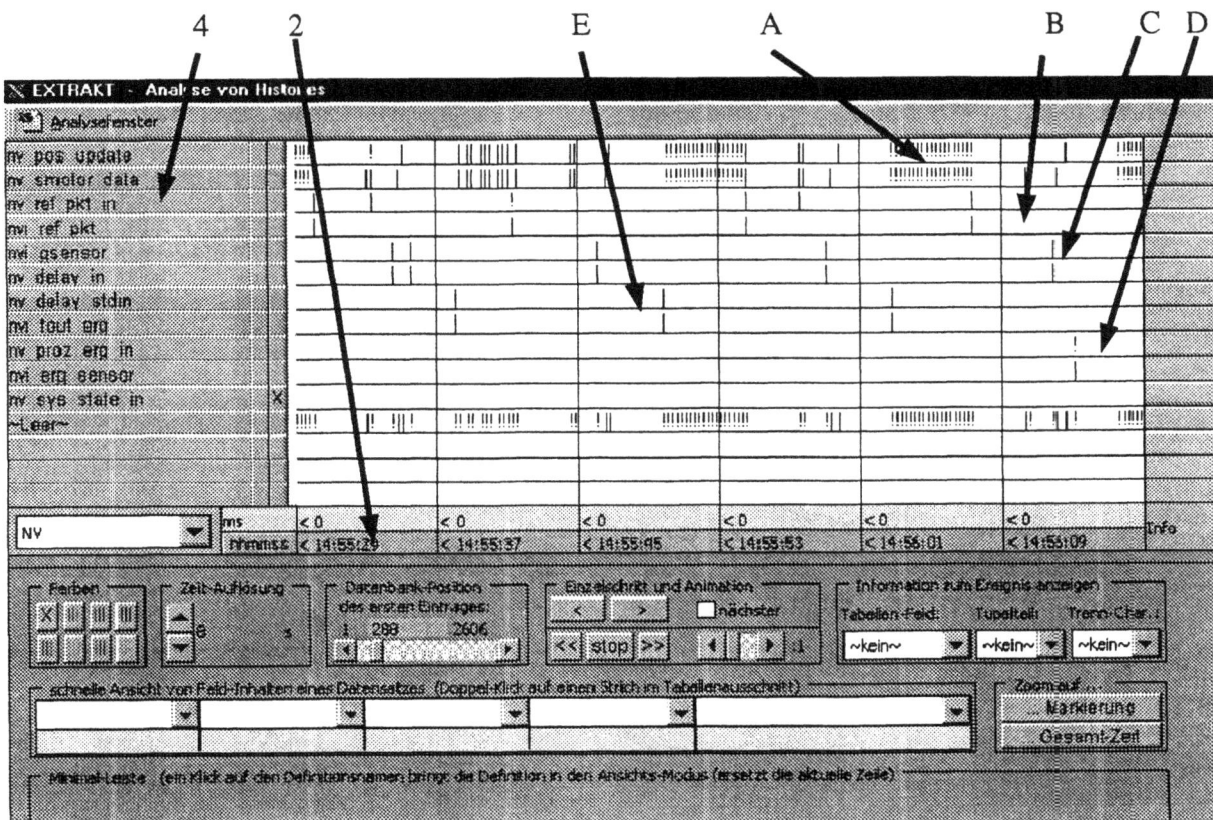

Fig. 3. The time analysis window shows the traffic at the network in the transport system
(list of values (detail 4), time (detail 2), details A, B, C, D, E see figure 2)

event
„Einlochen beginnt"
(transport begins)

event
„Kugel rollt über Arm"
(bead rolls)

event
„Kugel im Loch"
(bead in the hole)

variable
nv_ref_pkt_in
from
„Sensorik"
(sensors)
to
„Ablaufsteuerung"
(control of operation)

variable
nvi_gsensor
from
„Sensorik"
(sensors)
to
„Diagnose"
(diagnosis)

variable
nv_proz_erg_in
from
„Sensorik"
(sensors)
to
„Ablaufsteuerung"
(control of operation)

t_1 t_2 t_3 time

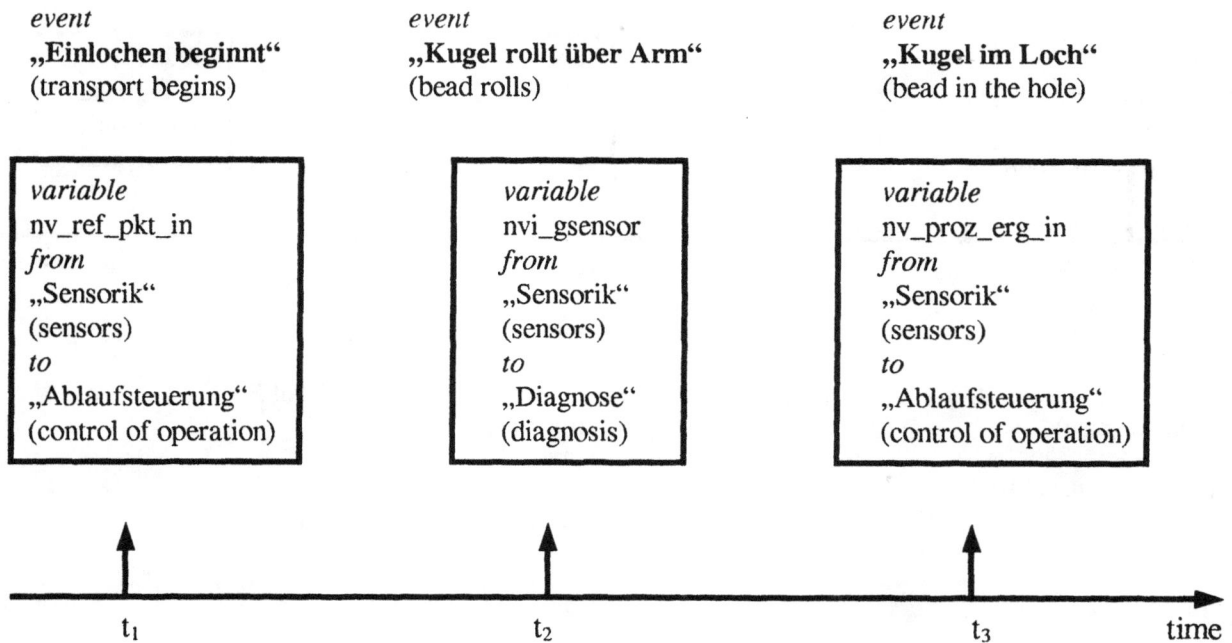

Fig. 4. The performance - pattern "Erfolgreiches Einlochen" (successful transport) is a sequence of three events (t_1, t_2, t_3)

X EXTRAKT - Analyse von Histories

Tabellenfenster Diagnose Filter

Zaehler	Zeit	ms	Service	Absender	Empfaenger	NV	DataNum	DataAsc	Einl	Kug	Kug	Erfo	Vorber
937	29.09.1998 14:56:07	322	ACK	Ablaufsteuerung	Schrittmotor	~Leer~	0	~Leer~					A
938	29.09.1998 14:56:07	390	Ackd	Sensorik	Ablaufsteuerung	nv_ref_pkt_in	1	1	E_				A
939	29.09.1998 14:56:07	399	ACK	Ablaufsteuerung	Sensorik	~Leer~	0	~Leer~					A
940	29.09.1998 14:56:07	405	Ackd	Sensorik	Diagnose	nvi_ref_pkt	1	1					A
941	29.09.1998 14:56:07	412	ACK	Diagnose	Sensorik	~Leer~	0	~Leer~					A
942	29.09.1998 14:56:07	419	Ackd	Ablaufsteuerung	Schrittmotor	nv_smotor_data	1	01 01 00 1e					A
943	29.09.1998 14:56:07	425	ACK	Schrittmotor	Ablaufsteuerung	~Leer~	0	~Leer~					A
944	29.09.1998 14:56:07	435	Ackd	Schrittmotor	Ablaufsteuerung	nv_pos_update	1	1					A
945	29.09.1998 14:56:07	443	ACK	Ablaufsteuerung	Schrittmotor	~Leer~	0	~Leer~					A
946	29.09.1998 14:56:10	464	Ackd	Ablaufsteuerung	Schrittmotor	nv_smotor_data	1	01 03 00 1e					A
947	29.09.1998 14:56:10	472	ACK	Schrittmotor	Ablaufsteuerung	~Leer~	0	~Leer~					A
948	29.09.1998 14:56:10	525	Ackd	Sensorik	Ablaufsteuerung	nv_ref_pkt_in	0	0	E_			R	A_
949	29.09.1998 14:56:10	532	ACK	Ablaufsteuerung	Sensorik	~Leer~	0	~Leer~					
950	29.09.1998 14:56:10	543	Ackd	Schrittmotor	Ablaufsteuerung	nv_pos_update	1	1					
951	29.09.1998 14:56:10	551	ACK	Ablaufsteuerung	Schrittmotor	~Leer~	0	~Leer~					
952	29.09.1998 14:56:10	673	Ackd	Ablaufsteuerung	Schrittmotor	nv_smotor_data	0	ff 01 00 1e					
953	29.09.1998 14:56:10	679	ACK	Schrittmotor	Ablaufsteuerung	~Leer~	0	~Leer~					
954	29.09.1998 14:56:10	691	Ackd	Schrittmotor	Ablaufsteuerung	nv_pos_update	1	1					
955	29.09.1998 14:56:10	698	ACK	Ablaufsteuerung	Schrittmotor	~Leer~	0	~Leer~					
956	29.09.1998 14:56:11	886	Ackd	Sensorik	Ablaufsteuerung	nv_delay_in	250	250					
957	29.09.1998 14:56:11	893	ACK	Ablaufsteuerung	Sensorik	~Leer~	0	~Leer~					
958	29.09.1998 14:56:11	902	Ackd	Sensorik	Diagnose	nvi_gsensor	1	1		E_	R		
959	29.09.1998 14:56:11	909	ACK	Diagnose	Sensorik	~Leer~	0	~Leer~					
960	29.09.1998 14:56:12	160	Ackd	Ablaufsteuerung	Schrittmotor	nv_smotor_data	0	ff 0c 00 1e					
961	29.09.1998 14:56:12	166	ACK	Schrittmotor	Ablaufsteuerung	~Leer~	0	~Leer~					
962	29.09.1998 14:56:12	509	Ackd	Schrittmotor	Ablaufsteuerung	nv_pos_update	1	1					
963	29.09.1998 14:56:12	516	ACK	Ablaufsteuerung	Schrittmotor	~Leer~	0	~Leer~					
964	29.09.1998 14:56:13	7	Ackd	Sensorik	Ablaufsteuerung	nv_proz_erg_in	1	1		E	R_		A
965	29.09.1998 14:56:13	14	ACK	Ablaufsteuerung	Sensorik	~Leer~	0	~Leer~					A
966	29.09.1998 14:56:13	21	Ackd	Sensorik	Diagnose	nvi_erg_sensor	1	1					A
967	29.09.1998 14:56:13	29	ACK	Diagnose	Sensorik	~Leer~	0	~Leer~					A

Fig. 5. Tables - view with events (E) and sequences (R)

32

Copyright © IFAC Distributed Computer Control Systems,
Sydney, Australia, 2000

ACRuDA

Odd Nordland

SINTEF Telecom and Informatics
Systems Engineering and Telematics
NO-7465 Trondheim

E-mail: Odd.Nordland@informatics.sintef.no
URL: www.informatics.sintef.no/~nordland

Abstract: Rail traffic is controlled by a set of interlocking computers distributed within the railway network and on-board computers on the trains. The safety assessment and validation of such computers shall not depend on the system architecture, so common assessment criteria are needed. The ACRuDA project consisted of identifying the state of the art, defining a common method and common criteria for assessment, applying them in case studies and refining them in the light of the case study experiences. The project succeeded in defining a common set of criteria that are applicable to all currently implemented digital architectures. *Copyright © 2000 IFAC*

Keywords: Distributed computer control systems, Railways, Safety

1. INTRODUCTION

Modern railway networks make extensive use of computer systems to control and monitor rail traffic in a safe and reliable way. A central element of such systems is the set of interlocking computers that control traffic and trackside equipment at stations and junctions. Here the term interlocking refers to the fact that any one computer must set the signals, points etc. that it controls in dependency of the settings at neighbouring locations. Each computer must therefore receive information from the neighbouring computers about the status of e.g. the signals and point switches that those computers control. In addition, it must get information about approaching and departing trains, and process this information together with its local data in order to set its own signals, point switches etc., and transmit appropriate data to the neighbouring systems for processing there. In addition, trains travelling in the area controlled by an individual interlocking computer must also receive up to date information.

On such networks, trains are also equipped with on-board computer systems that send data to and receive and process data from the interlocking computers. The kind of information sent to the train includes data about the train's maximum permissible speed, the distance to the next signal, the status of that signal etc. The on-board computer transmits data back to the interlocking computers, such as the train's identification, position, momentary speed and travelling direction etc.

Finally, there are rail traffic control centres that communicate with the interlocking computers to monitor and control traffic throughout the railway network, or at least a significant portion of it. If necessary, in suitably equipped networks traffic control centres can also intervene and stop trains directly.

Thus, safe rail traffic is achieved with a highly distributed set of computers, both static and mobile, performing dedicated portions of the overall control task.

Clearly, the safety of a modern railway network is then highly dependent on the reliability and safety of all these computer systems and their interactions. Therefore it is natural to subject such systems to a thorough safety assessment before authorising their use. However, this task is complicated by the fact that there are different system architectures that can be used to achieve a safe system.

A safety assessment should concentrate on the functions that the systems are to perform and not on the internal structure of the system performing those functions. In addition, safety assessments of differently structured systems should be comparable with each other, so that also the differently structured systems can be compared with each other. This gave rise to a need to identify assessment methods and criteria that were not dependent on the system's architecture.

2. ACRUDA

ACRuDA, "Assessment and Certification Rules for Digital Architectures" is a project that was partially funded by the Commission of the European Communities (CEC) under the Framework IV research program. The aim of the project was:

- to develop the assessment criteria for safety architectures used in the guided public transport industry,
- to develop a mutually recognised certification scheme for safety architectures in Europe, particularly in the fields:
    - numerical assurance (data redundancy)
    - redundant computer systems with voting mechanisms
    - distributed architectures
- to gain recognition from the European railway community,
- to establish a framework for developing safety cases,
- to ensure consistency with other related European projects (e.g. CASCADE).

The scope of the work encompassed both the application level and the computer platform level.

Participants were recruited from suppliers, operators and assessment bodies in various European countries. The suppliers were:

- ANSALDO Segnalemento Ferrovario,   Italy
- MATRA Transport International,   France
- ALCATEL Österreich,   Austria (represented by the Austrian Research Centre SEIBERSDORF).

Operators were:

- The Paris Metro operator RATP
- The Spanish national railways RENFE
- The French national railways SNCF.

Finally, the following assessor organisations participated:

- INRETS   France
- LLOYD'S Register   England
- SINTEF   Norway
- TÜV Rheinland   Germany.

The project was led by MATRA and divided into six work packages, under the leadership of different participants. Within each work package, tasks were defined and task leadership was allocated to various other participants, so responsibilities were fairly evenly distributed in the project. In the following sections, the task leaders are identified in brackets in front the task description.

3. WP1: STATE OF THE ART

This work package was under the leadership of INRETS. The objective was to gather all relevant information on current safety architecture designs and assessment methods. The following activities were defined:

- (INRETS) Examine safety critical architecture designs:
    - collect the design documentation of the current safety critical architectures and analyse the properties they claim to ensure
    - generate an initial list of design characteristics
- (SINTEF) Express current end user's requirements and their trends:
    - identify and analyse requirements from relevant standards
    - generate a list of applicable requirements

- (INRETS) Describe assessments of safety critical systems according to existing standards:
  - identify standards related to architecture safety (not only rail related)
  - determine commonalties
- (ANSALDO) Describe assessments of safety critical systems according to assessment methods:
  - identify methods that are currently applied in the certification process
  - determine a commonalties
- (SINTEF) Identification of lacks in standards:
  - investigate previously identified lacks in standards and practices
  - determine uncertainties and inapplicabilities of the standards

The work resulted in two reports describing the state of the art that were an excellent starting point for the next work package.

The first report, deliverable D1, provides a description of the currently used safety critical digital architectures and a survey of the state of the art in assessing the safety of such systems. It concluded that European train control equipment could integrate various types of safety critical architectures used in different countries.

The second report, deliverable D2, was more concerned with the use of standards and their application in various countries. The standards can be viewed as a collection of requirements that developers have to fulfil. The report provides recommendations on how to cope with unclear requirements, gaps in the standards and how to determine suitable measures.

4.   WP2: ASSESSMENT METHOD AND PROCEDURES

This work package, under the leadership of MATRA, aimed at defining a common method for validation and certification by analysing the methods and tools that had been identified in the previous work package. It contained the following tasks:

- (INRETS) Analyse safety plans, find commonalties:
  - select the most precisely described certification processes
  - analyse the relevance of the process according to the type of architecture
  - determine the commonalties, justify differences
- (ANSALDO) Define a common safety case structure.
- (MATRA) Define the validation method:

  - for each architecture list the technical points that are mandatory for safety
  - determine a set of common criteria
- (SNCF) Define a common certification method:
  - for each architecture list the procedures for certification
  - determine common and architecture specific procedures
- (SINTEF) Examine software certification procedures in other European projects:
  - compare the certification scheme with those defined in other European projects
  - propose a global certification scheme
- (MATRA) Define quality assurance consistent with ISO 9000 in certification and validation:
  - identify process assessment standards related to safety critical architecture development
  - collect process quality assurance guidelines

The work resulted in a report, deliverable D3, describing a mutually agreed method for assessing the safety properties of digital architectures and an agreed set of criteria to be applied. It provides information on the way a product or process is to be assessed by a third party and identifies top level criteria from which an assessor can derive detailed architecture specific criteria to be fulfilled.

It starts with a brief description of the background in terms of the certification requirements that result from the relevant EU directives. Then the principle concepts involved in the assessment and certification process are elaborated. The development process for a product is described, mainly based on the description given in the relevant CENELEC standards. This is the framework for the assessment and certification processes.

The recommended assessment and certification processes are then elaborated. This covers

- a description of the recommended assessment and certification processes,
- the roles of the different bodies (EU, state authorities, assessors etc.),
- the phases of the assessment process,
- re-use of assessment results,
- capitalisation of assessment work,
- certification report and certificate,
- assessment inputs,
- quality requirements for assessment activities,
- general concepts for assessment and certification of software.

Finally, the assessment criteria and the terminology round off the report. Deliverable D3 is the key result of the project.

5. WP3: TEST APPLICATIONS AND EXAMPLES

SNCF was responsible for this work package. Its objective was to determine how effective the assessment process that had been defined in the previous work package actually was by applying it to real cases. The tasks performed were:

- (SNCF) Choice of relevant situations to be assessed in safety studies:
  - define relevant assessment scenarios
  - choose suitable scenarios to be studied
- (RATP) Application of the validation cycle to a fail-safe architecture:
  - examine representative functions from previous task
  - perform a limited assessment in two different contexts
- (SNCF) Application of the assessment cycle to a European architecture.
- (SINTEF/Seibersdorf) Apply validation to a supplementary application.

Three digital architectures were selected:

- DIGISAFE, a single channel architecture using information redundancy (so-called coded monoprocessor). In this system, data is coded with supplementary information (e.g. time stamps) so that the software can detect any malfunction of the system.
- SARA, a multi-channel architecture using hardware redundancy. The channels have the same functionality, and a fail-safe voter is used to ensure correct functionality of the system.
- ELEKTRA, a multi-channel architecture using hard- and software diversity. In this system, one channel is used to monitor and control the other. This means that the software in each channel will be completely different. In addition, the channels use different hardware platforms, so that the system has diversity in both hard and software.

The criteria and methods that had been identified in the previous work package were applied to selected portions of DIGISAFE and SARA in order to demonstrate the methods' effectiveness.

This exercise revealed some shortcomings in the processes and criteria that had been defined. Applying the processes revealed that a considerable amount of supplementary information is necessary in order to perform a complete assessment. Deliverable D3 could not be used as a standalone document, but it was evidently a good place to start.

For ELEKTRA, a more global approach was adopted. Instead of performing an assessment on a limited portion of the system, all of the criteria were examined to determine if they were applicable to at least some part of the system.

This revealed certain weaknesses in the criteria, such as the use of qualitative expressions that required interpretation. There were also criteria that were phrased too concisely, assuming additional information (such as context, environment) that was not explicitly mentioned. Such criteria could be misunderstood.

Finally, some missing criteria were identified. These were criteria that were so "obvious" that nobody had thought of mentioning them, criteria that were implicitly included but not elaborated in detail or they were criteria that had indeed been overlooked, such as for example criteria for assessing validation and plans.

Some of these shortcomings could be rectified in the final version of deliverable D3, but unfortunately not all. Time and budget constraints left some questions unresolved.

6. WP4: EXPLOITATION AND DISSEMINATION.

The objective of this work package was to gain recognition for the certification scheme from the railway community. To this end, a Users' Group, consisting of about fifty representatives from suppliers and operators throughout Europe, was convened annually to discuss the results. This not only provided valuable feedback to the project, but also contributed to dissemination of the results of the project.

In addition, the project participants presented the project in a variety of fora and papers. This work package, under the leadership of Lloyd's register, is summarised in deliverable D6.

7. WP5: SYNTHESIS AND CONCLUSION

The work packages to date had produced a number of individual documents describing various aspects of the certification and assessment process. TÜV Rheinland was responsible for the synthesis of these documents into a single document, deliverable D7, which covered those aspects that were not contained in other deliverables. It provides background knowledge that is necessary to understand the functionality, design and safety properties of different types of architectures.

8. WP6: PROJECT MANAGEMENT.

This work package was the "bracket" that held everything together. MATRA Transport International had the overall responsibility for the project. The tasks, all of which were performed by MATRA, were:

- administration:
  - produce a project plan describing deliverables, activities
  - financial steering
- technical co-ordination:
  - identify technical problems
  - ensure that participants get the information they need
  - assess the impact of each task on project objectives
- quality assurance:
  - produce a quality assurance plan
  - control and monitor its application

The deliverables D8 and D9 are the visible results of this work package.

9. RESULTS

The principle results of the ACRuDA project are contained in a total of eleven deliverables, which are:

- D1: State of the Art:
  Safety Architectures Synthesis

- D2: State of the Art:
  Method Synthesis

- D3: The proposed assessment and certification methodology for digital architectures.

- D4: Assessment report - DIGISAFE

- D5: Assessment report - SARA

- D6: Dissemination report

- D7: Synthesis and conclusion

- D8: Project Plan

- D9: Quality Assurance Plan

- *D10: Code of Practice*

- D11: Test applications and examples; ACRuDA Assessment Results for ELEKTRA

Deliverables D1, D2 and D3 are public documents that can be obtained through the CEC or from the participants. They are also available on the Internet at www.informatics.sintef.no/~nordland/ACRuDA/ as HTML files. D10 does not exist as a separate document; it was incorporated into D3. Deliverables D6, D8 and D9 are of little interest outside the project, because they deal mainly with project management matters. They are restricted documents that may only be made available to the participants. The remaining reports D4, D5 and D11 contain proprietary information that is not freely accessible without express consent from the affected companies.

10. CONCLUSION

Deliverables D1 and D2 give a description of the state of the art at the time they were written (see also section 3 of this paper). Things have not changed substantially since then, so they will remain useful for some time to come. Deliverable D3 is a good reference to use when performing an assessment, although it's not a stand-alone document: supplementary information is necessary.

Nevertheless, the ACRuDA project has succeeded in defining a common set of criteria that are valid and applicable for all digital architectures that are currently implemented. The publicly accessible results of the project are a good starting point for performing an assessment and give a comprehensive survey of the formidable task of determining the safety characteristics of a microprocessor based safety system.

The ACRuDA project brought together suppliers, operators and assessors from a number of European countries. The exchange of knowledge and experience that this process produced was in itself a justification for the project. The participants would have liked to "polish" the results of the project and in particular resolve remaining open questions. Time and budget constraints prevented this. So there is still work left for an eventual follow on project!

Copyright © IFAC Distributed Computer Control Systems,
Sydney, Australia, 2000

SYSTEMS ENGINEERING OF TIME-TRIGGERED ARCHITECTURES – THE SETTA APPROACH[+]

C. Scheidler*, P. Puschner**, S. Boutin***, E. Fuchs[+], G. Gruensteidl[++], Y. Papadopoulos[+++], M. Pisecky[#], J. Rennhack[##], U. Virnich[###]

*DaimlerChrysler, **TU Vienna, ***Renault, [+]Dependable Computer Systems, [++]Alcatel Austria, [+++]University of York, [#]TTTech, [##]EADS Airbus, [###]Siemens

Abstract: SETTA addresses the systems engineering of safety-critical distributed real-time systems with a special focus on time-triggered architectures. An innovative methodology and a corresponding engineering environment is developed which aims for a higher maturity at early development steps. Key features are the support for a virtual systems integration and the tighter interconnection between the functional development process and the safety analysis process. The supporting tool components are designed and implemented in the course of the SETTA project. The methodology is evaluated by pilot applications from the automotive, aerospace, and railway domain. *Copyright © 2000 IFAC*

Keywords: Systems Engineering, Distributed Computer Control Systems, Real-Time Systems, Safety-Critical Systems, Fault-Tolerant Systems, Verification, Validation, Automotive Control, Aerospace Control, Railways.

1. INTRODUCTION

The overall goal of the SETTA project is to push the time-triggered architecture - an innovative European technology for safety-critical distributed real-time applications such as fly-by-wire or drive-by-wire - into future vehicles, aircraft, and train systems. To achieve this goal, SETTA focuses on the systems engineering of time-triggered architectures. The key characteristic of time-triggered systems is that all significant events, including tasks and messages, have to adhere to a pre-determined schedule (Kopetz and Gruensteidl, 1994). Time-triggered systems have several advantages, such as predictability concerning their real-time behavior, which make them uniquely suited for complex, safety-critical real-time systems (Scheidler, et al., 1997).

The paper is structured as follows. In section 2, the drawbacks in the systems engineering of time-triggered systems are analyzed. The SETTA systems engineering methodology, which overcomes these drawbacks, is proposed. The supporting tool components required for the SETTA methodology are identified. In section 3, the tool components developed in the course of the SETTA project are described in more detail. Section 4 covers the evaluation of the engineering methodology based on pilot applications from the automotive, aerospace, and railway domain. Section 5 concludes the paper.

[+] This work has been supported by the European Commission under IST contract number 10043. The SETTA consortium consists of the following partners: DaimlerChrysler, Alcatel Austria, EADS Airbus, Dependable Computer Systems, Renault, Siemens, TTTech, University of York, TU Vienna.

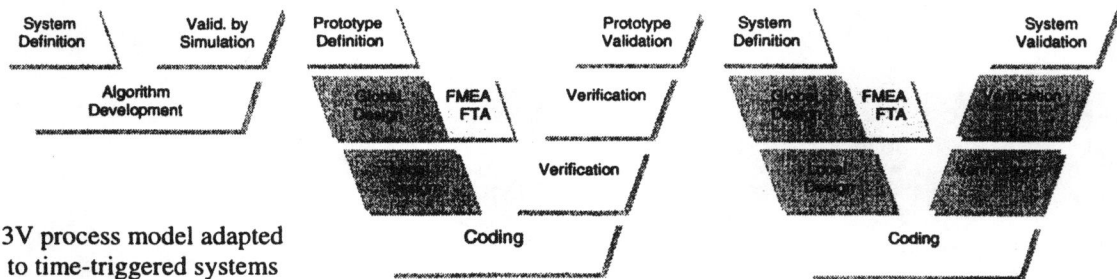

Fig.1: 3V process model adapted
to time-triggered systems

2. THE SETTA METHODOLOGY

2.1 Current drawbacks

Figure 1 shows the 3V lifecycle-process model which will be used to illustrate the weaknesses in engineering of time-triggered systems. The original 3V model has been firstly published by Mosnier and Bortolazzi (1997). The 3V model in *Figure 1* has been slightly adapted. Phases which traditionally put a major focus on the time-triggered nature of the target system are colored in dark gray.

The 3V model consists of three Vs representing the system simulation, prototyping, and product development stages.

The *first V* covers the definition and simulation of the overall system functionality. Software-in-the-loop-simulation (SIL) is the primary methodology applied. Implementation aspects, including the "time-triggered" property of the target system, are not considered in this systems-engineering phase.

The *second V* is characterized by rapid prototyping based on Hardware-in-the-loop simulation (HIL). In this phase, hardware specific parameters become important. The global design covers the mapping of tasks to computer nodes and the determination of the message scheduling between the nodes. The local design addresses the scheduling of tasks on each node.

The *third V* addresses the system development for the final target hardware. A typical problem at this stage is the limited performance of the target system. Deadlines met by the oversized prototypical hardware might not be met on the target -- a situation that is not acceptable for the safety-critical systems targeted at in the SETTA project.

At least four drawbacks in this lifecycle-process model can be identified.

1. *There is a gap between the first and the second V.* Due to constraints of the target system, the assignment process of the second V may fail, thus invalidating the result of the preceding simulation stage of the first V. For example, a distrib-

uted control application running stable at the first V might behave differently due to the timing constraints caused by message passing between computer nodes.

2. *A schedule verification tool on the global design level is lacking.* The verification tool is needed to check the consistency of the message descriptor list (MEDL; the MEDL determines the message schedule and thus the run-time behavior of the final system). A verification tool is particularly demanded for the acceptance of the time-triggered technology in the aerospace industry.

3. *A timing verification tool on the local design level is lacking.* Executing code and measuring its execution time on the target is the current state-of-the-art. However, this technique cannot guarantee to yield safe upper bounds of the execution time.

4. *The functional development process and the safety analysis process are de-coupled.* Tools supporting Fault-Tree Analysis (FTA), Event-Tree Analysis (ETA) and Failure Mode and Effects Analysis (FMEA) are not connected to simulation tools like Matlab/Simulink.

2.2 SETTA Design Flow Model

The aim of the SETTA project is to propose a design-flow model for time-triggered systems that overcomes the four shortcomings identified before.

A key component in the SETTA approach is a suite of simulation-building blocks provided for the Matlab/Simulink environment. The simulation building blocks support a virtual systems integration, in other words, the gap between the first V and the second V is closed. Time-triggered systems are, in contrast to event-triggered systems, fully predictable in their run-time behavior. SETTA exploits this predictability at the modeling stage. Simulation-building blocks model not only the core functionality of a system, but also the distributed nature and the used communication mechanisms, which both affect the system's behavior. E.g., the effects of value discretisation, communication delays, and fault-tolerance, which are

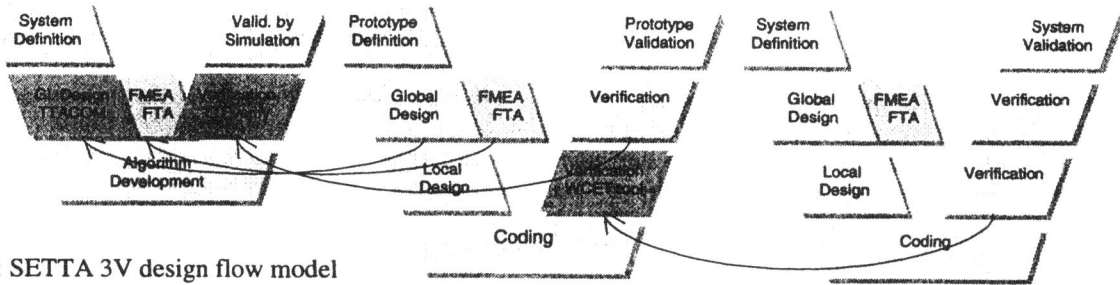

Fig. 2: SETTA 3V design flow model

typically abstracted in a system model and are a significant source of problems in later implementation stages, are much easier dealt with, if they are already part of the system model. Based on the virtual systems integration, system manufacturers and component suppliers can co-operate in a much tighter way.

A schedule verification tool for the global design level is developed. Verification on the global design level, nowadays performed at the prototyping or system development stage, is mapped to the simulation stage. The verification tool developed within the SETTA project verifies the MEDL which controls the simulation building blocks.

A timing verification tool for the local design level is developed. A specific WCET tool for the simulation-building block set analyses the timing behavior of the code generated for each of the building blocks.

An algorithm for fault-tree synthesis will be developed and implemented which provides an intelligent interface between a functional modeling tool (Matlab/Simulink from The Mathworks) and a fault tree analysis tool (FT+ from Isograph).

To summarize, the main goal is to achieve a high maturity at early development stages based on virtual systems integration. Activities nowadays performed at prototyping or product development stage are mapped to the simulation stage, as depicted by the arrows in *Figure 2*.

Although the tool components are developed by four different partners, SETTA aims for an fully integrated systems engineering environment. To achieve this goal, the tool components will be linked via different interfaces which will be sketched in the following.

The simulation building blocks are triggered by a configuration file which describes the message scheduling on the bus interconnecting computer nodes. This file can be checked by the scheduling verification tool. The WCET analysis will be integrated into the Matlab/Simulink environment and is therefore also connected to the simulation building blocks. The fault-tree synthesis algorithm extracts the

structural information out of Matlab/Simulink files which can be extended with the simulation building blocks.

3. SETTA TOOL COMPONENTS

3.1 Simulation Building Blocks

A suite of simulation-building blocks - product name: TTACOM - is developed in the SETTA project which support the virtual systems integration (involved partner: Dependable Computer Systems). TTACOM is a Matlab/Simulink blockset that allows the development of distributed applications, including the Time-Triggered Protocol (TTP) bus. It contains blocks for configuring clusters, reading and writing messages, controlling the simulation progress over time, and a detailed graphical TTP interface.

The blockset provides the application designer with a realistic interface to the TTP communication layer, without going to real hardware. Especially the timing and the fault tolerance mechanisms of the underlying communication system can be explored in detail. The main objective of TTACOM is to model the behavior of a TTP cluster as accurate as needed for application design.

Due to the distributed nature of TTP, the blockset is designed to support the modeling of distributed applications. The static nature of TTP allows the precise modeling of communication latencies given by the global communication schedule (Message Descriptor List MEDL). Therefore, the simulation of a distributed control algorithm considers also real-world communication latencies.

TTP provides a number of fault tolerance mechanisms which can be exploited for safety critical applications. To validate the fault tolerance properties of the application, TTACOM allows for injection of faults in the communication layer. Simulation based fault injection is a powerful technique to analyze failure scenarios. In contrast to real hardware, fault injection in simulation models can be done more systematically and with less effort.

3.2 Schedule Verification

A schedule verification tool - product name: TTPverify - is developed in the SETTA project (involved partner: TTTech). The purpose of this component is to check the message descriptor list (MEDL) of time-triggered systems.

The communication in time-triggered systems is statically scheduled. The communication controllers transmit data according to a predefined schedule which is stored in dedicated data tables. To ensure correct system functionality, it is therefore of the utmost importance to verify that these automatically generated data tables satisfy the overall requirements. For this purpose, a dedicated schedule verification tool will be specified which can read the data tables and verify that they meet the requirements.

MEDL files are especially checked vs.
- TTP Specification
- specification of the TTP controller
- consistency between different MEDL files of the same cluster
- completeness of a set of MEDL files

The verification tool will be certified according to aerospace standards.

3.3 WCET Analysis

The WCET analysis tool that is being developed in the course of the SETTA project (Kirner 2000) derives the WCET by means of static code analysis (involved partner: Technical University of Vienna). This stands in contrast to the widely used method of determining the WCET by measuring the duration of representative task executions. This latter approach cannot provide a guaranteed execution- time bound.

At the beginning of the WCET analysis tool chain, a Matlab/Simulink model of the application is used to generate the relevant processing tasks for the real-time system. To enable easy transfer of the processing tasks to the target hardware and maintain a high degree of readability, the program code for each task is produced in C. Using C leads to a generic and portable representation.

The execution time of the processing tasks is largely determined by the temporal execution properties of the CPU. Therefore, it is reasonable to perform the WCET analysis using the machine code representation of the processing task. Only on this level it is possible to apply the knowledge of the execution time of each machine instruction. Moreover, the analysis on the machine code level allows us to consider potential optimizations performed by the compiler.

The quality of the result depends on how far the WCET bound determined by the analysis tool over-estimates the actual WCET. It is not possible to derive high quality bounds on the execution time for every type of code: Restrictions, such as avoiding indeterminate loop count, must be considered during code generation to facilitate WCET.

The quality can be significantly improved, if the WCET analysis tool is provided with more information on program flow than just the program-flow path implied in the C code. Examples for such additional information are a maximum loop count or the identification of mutually exclusive program paths. Most often this information can be derived from the application model. Hence a way must be found to convey it to the WCET tool. In our approach, the C-code and the machine-code representation of the processing task are complemented with appropriate meta-information. The results of the WCET analysis are back-annotated to the C-code representation and the Matlab/Simulink model, which facilitates examination of the results and well directed code optimization.

3.4 Safety Analysis

The work on safety analysis (involved partners: University of York) will continue based on the results of the SETTA predecessor project TTA (Papadopoulos 2000). One of the main results of this project was a new method for safety analysis which potentially integrates the design and safety analysis of time-triggered architectures and simplifies the assessment by enabling some degree of mechanical safety analysis during the synthesis of fault trees for the system.

The aim of the SETTA project is to improve the design and robustness of the method, so that it can be applied in a more realistic industrial context. It is intended to implement the fault tree synthesis algorithm as a direct interface between a widely used functional modeling tool (Matlab/Simulink from The Mathworks) and a similarly popular fault tree analysis tool (FT+ from Isograph).

The synthesis algorithm performs a backward traversal of the data flow graphs given by the Matlab/Simulink system model. The algorithm starts at an output port which is associated with a top event of a fault tree. It then traverses the data flow graph and identifies all components which can cause the top event. The algorithm assumes that the output of a component is deviated if the component itself has a failure or the inputs of the components are deviated. Relevant failure classes are omission, commission and value failures.

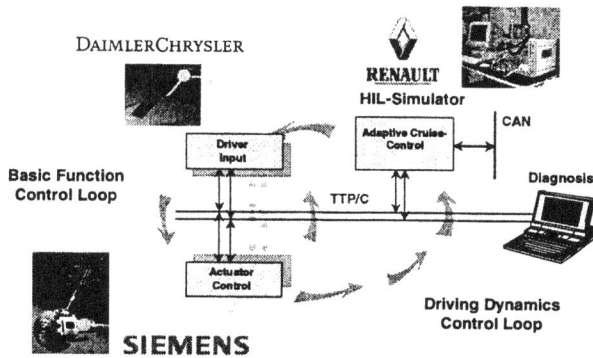

Fig. 3: Automotive Validator

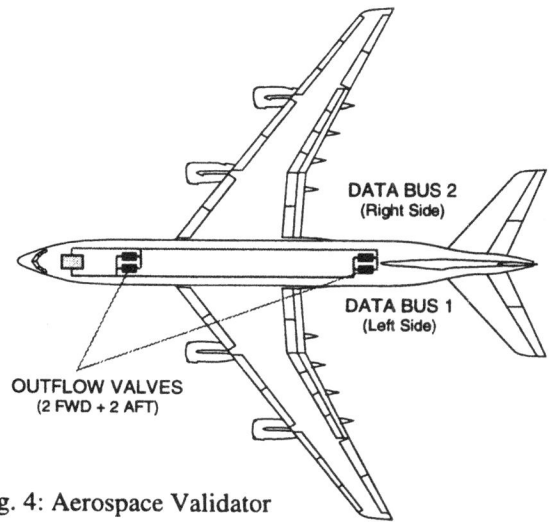

Fig. 4: Aerospace Validator

4. THE VALIDATORS

4.1 Automotive Validator

The objective of the automotive validator is to evaluate the results of the SETTA project in the automotive domain (involved partners: DaimlerChrysler, Renault, and Siemens Automotive). The architecture chosen to be the validator for SETTA is a part of an automotive chassis control system which consists of a brake-by-wire system and an adaptive cruise control simulator *(Figure 3)*. The brake-by-wire system consists of a redundant brake pedal system provided by DaimlerChrysler and a brake actuator provided by Siemens Automotive. The adaptive cruise control simulator provided by Renault models the dynamics of a vehicle on a highway. The system has strict performance, timing, and safety requirements and contains two distributed control loops, depicted by arrows in *Figure 3*. The development of the automotive validator will be performed in accordance with the SETTA design flow approach. All tool components developed in the project will be evaluated.

In particular, the concept of virtual systems integration will be evaluated in depth. The three automotive partners firstly develop system models of their components, which describe the normal and malfunction behavior of the component. The components are than virtually integrated via the simulation building blocks. Intensive tests can be performed before the physical system is built. This includes also extensive fault-injection experiments. A typical fault-injection scenario for an electronic braking system is the loss of a wheel node. The braking functionality can still be provided by the three remaining wheels nodes.

In the second step, the automotive validator is integrated physically. It is the vision of the automotive validator team to demonstrate that the SETTA environment supports "plug-and-play" for safety-critical distributed real-time systems.

4.2 Aerospace Validator

The objective of the automotive validator is to evaluate the results of the SETTA project in the aerospace domain (involved partner: EADS Airbus). The architecture chosen to be the validator for SETTA is the cabin pressure regulation system. This system has strict performance, timing and safety requirements. Two independent pressure control functions will be realized for backup reasons and will be implemented as redundant components. Both controller functions will receive appropriate information such as planned flight profile, current position, altitude, and current cabin pressure from the air data/inertial reference system. Taking these parameters and the actual cabin pressure into account, the pressure controller will calculate and command the desired openings for the outflow valves. All tool components developed in the project will be evaluated. The aerospace validator will be integrated virtually (first step) and physically (second step).

4.3 Railway Validator

In the SETTA project, Alcatel Austria provides the specific requirements from the railway domain and validates the SETTA engineering methodology and tools by using a typical railway application.

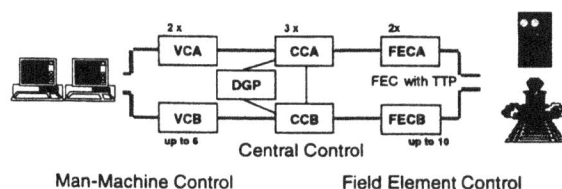

Fig. 5: Railway Validator

43

The selected application covers the field element controlling subsystem of Alcatel's Electronic Interlocking System ELEKTRA. The railway validator is more related to a real product, compared to the prototypical automotive and aerospace validator.

The Electronic Interlocking System ELEKTRA is designed with regard to function, safety and reliability, and is divided into three main units.

The task of the central control (CUA, CUB in *Figure 5*) unit is to verify and edit the interaction between elements or groups of elements as a result of inputs by the operator. Commands and internal transitions are generated to be transmitted to the external installation as well as to the operator. To be able to meet the high availability requirements, the central unit is designed in triplicate.

The operating unit is the man-machine interface (VC) between operator and central unit. By using mouse and keyboard, the operator sends commands to the central unit. The actual states of elements are displayed in real time.

The function of the field element control unit (FEC) is to convert commands generated by the central unit and to transmit them to field elements. Simultaneously, information about the conditions of the external installation is transferred to the central unit. Double redundancy is used for availability.

A specially designed diagnosis device saves all relevant information to allow efficient maintenance. To meet the high requirements regarding safety, a distinction is made between two software channels; the logic channel (A) and the safety channel (B). After having processed all inputs in the logic channel, the data is checked in the safety channel before transmitted to the external.

The main focus of the railway validator is the evaluation of the schedule verification and timing verification tool.

5. SUMMARY AND OUTLOOK

A new design methodology for the systems engineering of time-triggered systems is developed which aims for a higher maturity at early design steps and a better interconnection between the functional development process and the safety analysis process.

A key issue in the design methodology is the concept of virtual systems integration provided by a suite of Matlab/Simulink building blocks. These blocks support the modeling of the communication behavior of a distributed time-triggered systems. The configuration file driving the simulation can be checked by the schedule verification tool. A timing verification tool is developed which performs a worst case execution time analysis for pieces of "C" code. WCET analysis is needed to analyze the interaction of message and task scheduling. Finally, a fault tree synthesis tool is developed which generates fault trees out of simulation block diagrams. The tool components will be linked via interfaces and integrated to a systems engineering environment.

The methodology and the supporting environment are evaluated in the automotive, aerospace, and railway domain, which will demonstrate the feasibility of the methodology. In addition, first estimated cost saving figures are expected.

The SETTA consortiums hopes that their research activities make a substantial contribution to the systems engineering of time-triggered systems, which is expected to push this promising European technology into future vehicles, aircraft, and train systems.

6. ACKNOWLEDGEMENT

The author would like to thank all SETTA partners for their active support in the project. The author is grateful to Peter Puschner for writing the first version of this paper.

REFERENCES

Kirner, R., R. Lang, P. Puschner (2000). Integrating WCET Analysis into a Matlab/Simulink Simulation Model . Submitted for *DCCS 2000: 16th IFAC Workshop on Distributed Computer Control Systems,* Sydney, Australia, 29th November – 1st December.

Kopetz, H., G. Gruensteidl, (1994). TTP – A Protocol for Fault-Tolerant Real-Time Systems. *IEEE Computer*, **Vol. 24 (1)**, pp. 14-23.

Mosnier, F., J. Bortolazzi, (1997). Prototyping Car-Embedded Applications. In *Advances in Information Technologies: The Business Challenge*, pp. 744-751, IOS Press.

Papadopoulos Y., (2000) Mechanical Synthesis of Fault Trees from Hierarchical Design Models, Epigram Newsletter, Issue 9, In *CUIG, the European Core Interest Group in Programmable Safety Related Systems*, pp 1-2.

Scheidler, C., G. Heiner, R. Sasse, E. Fuchs, H. Kopetz, C. Temple. (1997). Time-Triggered Architecture – (TTA). In: *Advances in Information Technologies: The Business Challenge*, pp. 758-765. IOS Press.

Copyright © IFAC Distributed Computer Control Systems,
Sydney, Australia, 2000

CHECKPOINT PLACEMENT FOR FAULT-TOLERANT
REAL-TIME SYSTEMS

Hyosoon Lee, Heonshik Shin, Naehyuck Chang

School of Computer Science and Engineering
Seoul National University, Seoul 151-742, Korea
{fanta,shinhs,naehyuck}@snu.ac.kr
Phone: +82-2-880-7295
Fax: +82-2-874-3104

Abstract: The objective of this paper is to analyze the worst case timing requirement of real-time tasks with checkpointing and to provide the optimal checkpoint placement scheme for fault-tolerant real-time systems. The proposed scheme aims at minimizing the worst case timing requirement of a task which has up to n checkpoints to tolerate a finite permissible number of failures. Simulation result shows that, though it does not increase the average case timing requirement of a task too much, it reduces the worst case timing requirement of a task by about 7% to 26% compared with modified equidistant scheme. The timing information derived in this paper makes checkpointing applicable to real-time systems while keeping the validity of schedulability check developed in the past. *Copyright ©2000 IFAC*

Keywords: Fault tolerance, Reliability, Checkpointing, Real-time systems, Execution time

1. INTRODUCTION

Modern control applications have had an increasing demand for more complex and sophisticated real-time computing systems in such areas as nuclear plant control and factory automation. In particular, fault tolerance is one of the important requirements in the design of real-time distributed systems (Bertossi and Mancini, 1994). WCET (Worst Case Execution Time) of real-time tasks is an important metric that is commonly used to guarantee their schedulability. Unfortunately, it has been assumed that real-time tasks run without failure occurrence during the mission time. The execution time of a task must be defined as the time to complete computation, including fault treatment of a task. It is known that the expected execution time of a task with restart capability typically grows exponentially with its productive processing time[1] in

the presence of random faults. With checkpointing, the expected execution time of a task typically grows only linearly with its productive processing time. Although checkpointing is a commonly used technique in fault-tolerant applications to enhance their reliability and to reduce the extra time required for recovery process, restart has been used in real-time systems (Bertossi and Mancini, 1994; Ghosh, 1996) because checkpointing requires costly run-time overhead. However, current technology allows checkpoints to be performed as atomic operations with a very low overhead and at a very high rate, thus making the schemes proposed here attractive. For example, Bowen and Pradhan (1993) describe processor- and memory-based checkpointing schemes which allow checkpointing to be performed with negligible overhead, and the Spring kernel provides the abstraction of a fast global shared memory implemented by a fiber optic ring. Many distributed real-time systems also perform a synchronous fault-detection protocol every few milliseconds. In such systems, one could think of taking a synchronous checkpoint of the state of the tasks running on all the non-faulty processors at every

---

This work was financially supported by Agency for Defense Development and by ACRC(Automatic Control Research Center), Seoul National University.

[1] It is the execution time of a task excluding checkpointing overhead, time spent in repair and reprocessing time.

execution of the failure-detection protocol (Bertossi and Mancini, 1994). However, checkpointing may not always be profitable: establishment of more checkpoints leads to less reprocessing time after failures, but extra overhead for the establishment of checkpoints increases. Therefore efficient checkpoint placement schemes must be devised to maximize the checkpointing profit. Our objectives are to provide the optimal checkpoint placement scheme for fault-tolerant real time system and to make it applicable to well-known schedulability check methods. The scheme focuses on minimizing the worst case timing requirement of tasks to maximize the quantum of laxity in the presence of permissible number of failures.

The remaining part of the paper is organized into five sections. Section 2 introduces previous checkpoint placement schemes. Fault model and assumptions of our work are described in Section 3. Section 4 derives the optimal checkpoint placement scheme for real-time tasks which may suffer a finite permissible number of failures during the mission time. It also presents some results on how much WCET of tasks can be reduced by our scheme through simulation. Section 5 describes how to calculate a permissible number of failures. Finally, Section 6 provides concluding remarks.

2. CHECKPOINT PLACEMENT

Checkpointing was originally introduced to database systems in order to enhance its reliability. Upon a hardware or software failure, the system is restored to the last checkpoint. This avoids the need to re-process all execution after the failure, thus improving the system availability by reducing the average processing time. It is also very economical in comparison with hardware redundancy and restart capabilities. *Equidistant* and *equicost* are the representatives of checkpoint placement schemes. Equidistant scheme was shown to maximize the steady-state system availability, and to minimize the expected program execution time in the presence of random faults. Irrespective of failure distribution, all checkpoint intervals [2] are the same in this scheme. Therefore, it has often been considered by system analysts and designers because of its intuitive appeal and simplicity (Nicola, 1995). Shin *et al.* (1987) propose the checkpoint placement scheme which minimize the expected deadline miss probability and show that equidistant scheme is optimal in the basic model. In equicost scheme, a checkpoint is performed whenever the expected reprocessing cost is equal to the checkpointing cost. It was demonstrated that, for a Weibull failure distribution, equicost scheme achieves higher system availability than an equidistant scheme (Tantawi and Ruschitzka, 1984).

Plank *et al.* (1995) use a different approach, so called synchronous checkpointing that can improve on the performance of asynchronous techniques and can also substitute for them when automatic mechanisms are not available. This technique is a user directive that allows the programmer to specify points in the program where it is most advantageous for checkpointing to occur.

Roll-forward checkpointing is developed to avoid rollback to the previous checkpoints with hardware redundancy in addition to time redundancy (Pradhan and Vaidya, 1994; Long *et al.*, 1990). Since these approaches adopt equidistant scheme, the WCET of tasks with roll-forward checkpointing is the same as that of equidistant scheme in the worst case multiple failure occurrence.

Checkpointing has been well analyzed for tasks with finite productive processing time. Whereas the case of minimizing the expected execution time is the most common, probabilistic analysis of expected execution time of tasks in real-time systems may not be important because hard real-time tasks become valueless or disastrous on deadline miss. There are few approaches to analyze the behavior of checkpointing in real-time systems (Lee and Shin, 1999; Burns *et al.*, 1996; Punnekkat, 1997). They are, however, not only equivalent to equidistant scheme for the most part but also restricted by non-practical assumptions.

3. FAULT MODEL AND ASSUMPTIONS

This work treats hard real-time tasks running on the environment where a finite number of transient or intermittent failures can occur. It is deemed reasonable because previous studies (Ghosh, 1996; Iyer *et al.*, 1986; Castillo *et al.*, 1982) have indicated that transient errors account for majority of the faults observed during the life time of systems, and infinite time redundancy is required to tolerate unbounded number of failures. There are various models proposed to characterize the failure process. It is the most commonly used that the time between failures follows an exponential distribution, Weibull distribution or their variations (Castillo *et al.*, 1982). Although this analysis is applicable to other failure models just by substituting the corresponding failure distribution function for calculation, it is assumed that the time for a failure to occur is exponentially distributed with constant failure rate λ for simplicity. Thus, the probability distribution function is given by

$$Pr(t < T) = 1 - e^{-\lambda T}$$

The system is assumed to be augmented with some capabilities to detect faults. Usually task failures are detected by acceptance tests or watchdog timers which interrupt the execution of task. The cost and overhead of fault detection of a task τ_i is included in the checkpointing overhead C_i. It is assumed that an acceptance

---

[2] It is defined as the duration time between the establishment of two consecutive checkpoints. An interval begins when one checkpoint is established, and ends when the next checkpoint is established.

Table 1. List of symbols

Symbol	Meaning
R_i	Rollback overhead of τ_i
T_i	i^{th} checkpoint interval
f_i	i^{th} failure occurrence time
s_i	i^{th} checkpoint start time
p_i	The last checkpoint start time until f_i
P_i	Checkpoint interval which f_i is in
v_i	The first checkpoint start time after f_i
n_i	The last checkpoint index until f_i
b	The number of failures occurring within the last checkpoint interval

test is executed when checkpoints are established to avoid any latent faults. Thus, a fault affects only the currently executing tasks by virtue of fault containment and zero error latency. It is also assumed that checkpointing can be accomplished at any time within the mission time of a task. If a fault occurs, the task which is being currently executed will be returned to its last checkpoint. Execution will then proceed with a task having the same priority as before. All tasks are hard real-time tasks with a hard deadline. Productive processing time w_i of a task τ_i, which is generally denoted by WCET in ordinary real-time systems, can be determined *a priori* by repeated tests and validation of the task using timing tools. The timing requirement of a task to guarantee its timeliness must include the productive processing time, run-time overhead due to checkpoint establishment and reprocessing time due to failure recovery. Let $\alpha_i(n, k_i)$ denote the extra time required for checkpoint establishment and recovery process in the presence of k_i failures during the mission time of a task τ_i with $n + 1$ checkpoint intervals, and let $\mathcal{A}_i(n, k_i)$ denote the worst case value of $\alpha_i(n, k_i)$. Then, the worst case timing requirement $W_i(n, k_i)$ of τ_i to guarantee its feasibility can be calculated as follows:

$$W_i(n, k_i) = w_i + \mathcal{A}_i(n, k_i) \qquad (1)$$

This value will be valid as long as no more than the permissible number of failures are exhibited. Table 1 lists additional notations used in this paper.

4. STATIC CHECKPOINT PLACEMENT

This section refines Eq. (1) for the tasks adopting checkpointing, and infers the optimal checkpoint placement scheme which minimize it.

Lemma 1. The worst case timing requirement of a task τ_i with n checkpoints in the presence of no more than k_i failures is minimized when checkpoint intervals satisfy

$$T_1 = T_2 = \ldots = T_n = T_{n+1} - C_i$$

Proof Since w_i is the summation of all productive processing time between checkpoints and $n \cdot C_i$ is the

run-time overhead required for performing n checkpoints,

$$w_i = \sum_{j=1}^{n+1} T_j, \text{ and } \alpha_i(n, 0) = n \cdot C_i$$

$\alpha_i(n, j)$ can be defined recursively as follows:

$$\begin{cases} R_i + f_j + \alpha_i(n, j - 1), & \text{if } p_j = 0 \\ R_i + f_j - (p_j + C_i) + \alpha_i(n, j - 1), & \text{otherwise} \end{cases}$$

If l is the number of failures occurring in the first checkpoint interval, the range of f_j becomes

$$\begin{cases} p_j \leq f_j < v_j + C_i, & \text{if } 0 < j \leq l \\ p_j + C_i \leq f_j < v_j + C_i, & \text{if } l < j < k_i - b \\ p_j + C_i \leq f_j < v_j, & \text{if } k_i - b + 1 < j \leq k_i \end{cases}$$

Therefore, $\mathcal{A}_i(n, k_i)$ can be written as

$$\mathcal{A}_i(n, k_i) = k_i \cdot R_i + (k_i + n - b) \cdot C_i$$
$$+ \sum_{j=1}^{k_i} P_j \qquad (2)$$

Since $-b \cdot C_i + \sum_{j=1}^{k_i} P_j$ is the summation of k_i intervals repeatedly selected from $\{T_1, T_2, \ldots, T_n, T_{n+1} - C_i\}$, $\mathcal{A}_i(n, k_i)$ is minimized when $T_1 = T_2 = \ldots = T_n = T_{n+1} - C_i$. Thus, the value of T_i for $1 \leq i \leq n$ equals $(w_i - C_i) / (n + 1)$ and the value of T_{n+1} equals $(w_i - C_i) / (n + 1) + C_i$. $W_i(n, k_i)$ of this scheme can be obtained as follows:

$$W_i(n, k_i) = w_i + n \cdot C_i + k_i \cdot \left(C_i + R_i + \frac{w_i - C_i}{n + 1} \right)$$

□

Corollary 2. The optimal number of checkpoints n^* of a task in the presence of k_i failures is given by

$$n^* = \begin{cases} \lceil \sqrt{m} \rceil - 1, & \text{if } \lceil \sqrt{m} \rceil \cdot \lfloor \sqrt{m} \rfloor > m > 1 \\ \lfloor \sqrt{m} \rfloor - 1, & \text{else if } m > \lceil \sqrt{m} \rceil \cdot \lfloor \sqrt{m} \rfloor > 1 \\ 0, & \text{otherwise} \end{cases}$$

where $m = k_i \cdot (w_i - C_i) / C_i$.

Proof The optimal number of checkpoint n^* can be easily obtained directly from $\partial W_i(n, k_i) / \partial n = 0$ and $\partial^2 W_i(n, k_i) / \partial n^2 > 0$. □

If checkpointing is kept after the occurrence of the last tolerable failure, Lemma 1 shows that equidistant checkpoint placement is also optimal in this deterministic analysis except that the last checkpointing is not performed because of the completion of the task.

We now consider the checkpoint placement scheme where checkpoint intervals are adapted according to the last permissible failure occurrence time. Without difficulty, it can be found that it is unnecessary to keep performing checkpointing after the last permissible failure occurrence for a task.

Lemma 3. The worst case timing requirement of a task τ_i of which the maximum number of checkpoints is n to tolerate a single failure is minimized when checkpoint intervals satisfy the following:

$$T_1 = T_2 + C_i = \ldots = T_n + (n-1) \cdot C_i$$
$$= T_{n+1} + (n-1) \cdot C_i$$

Proof $\mathcal{A}_i(n,1)$ of a task τ_i can be formulated as follows:

$$\mathcal{A}_i(n,1) = R_i + C_i \cdot (1 + n_1 - b) + P_1 \qquad (3)$$

In the above equation, $C_i \cdot (n_1 - b) + P_1$ is the only variable part which is referred to as *var* here. Let $L_1, L_2, \ldots, L_n, L_{n+1}$ denote $T_1, T_2, \ldots, T_n, T_{n+1} - C_i$, respectively. Then,

$$var = C_i \cdot n_{max} + MAX_{j=1}^{n_1+1} L_j$$
$$- \{MAX_{j=1}^{n_1+1} L_j - L_{n_1+1} + C_i \cdot (n_{max} - n_1)\}$$

where n_{max} denotes the last checkpoint index until the checkpoint interval $MAX_{j=1}^{n_1+1} L_j$. Assume that there is a checkpoint placement scheme of which the first checkpoint interval is T_x. If T_{x+1} is greater than $T_x + C_i$, *var* has a greater value when the failure occurs within the second interval than within the first interval. The reason is that the value of $\{MAX_{j=1}^{n_1+1} L_j - L_{n_1+1} + C_i \cdot (n_{max} - n_1)\}$ becomes positive when a failure occurs in the first interval. Iteratively, if the condition of $L_{j+1} \leq L_j - C_i$ is satisfied for all $j \geq x$, the worst case timing requirement of task is not more than that when the failure has occurred before j^{th} checkpoint establishment. Thus, $W_i(n,1)$ is minimized when $T_2 = T_1 - C_i, T_3 = T_2 - C_i, \ldots, T_n = T_{n-1} - C_i, T_{n+1} = T_n$. Since $w_i = \sum_{j=1}^{n+1} T_j$, $T_1 = (w_i - C_i)/(n+1) + (n \cdot C_i)/2$ and hence

$$W_i(n,1) = w_i + R_i + C_i + \frac{w_i - C_i}{n+1} + \frac{n \cdot C_i}{2}$$

□

Corollary 4. The optimal number of checkpoints n^* of a task to tolerate a single failure is given by

$$n^* = \begin{cases} \lceil \sqrt{m} \rceil - 1, & \text{if } \lceil \sqrt{m} \rceil \cdot \lfloor \sqrt{m} \rfloor > m > 1 \\ \lfloor \sqrt{m} \rfloor - 1, & \text{else if } m > \lceil \sqrt{m} \rceil \cdot \lfloor \sqrt{m} \rfloor > 1 \\ 0, & \text{otherwise} \end{cases}$$

where $m = 2 \cdot (w_i - C_i)/C_i$.

Proof The optimal number of checkpoint n^* can be easily obtained directly from $\partial W_i(n, k_i)/\partial n = 0$ and $\partial^2 W_i(n, k_i)/\partial n^2 > 0$. □

Note that the checkpoint placement scheme according to Lemma 3 places more checkpoints towards the end of task as that of Shin *et al.* (1987). Based on Lemma 1 and Lemma 3, the optimal checkpoint placement scheme can be inferred as shown in the following theorem.

Theorem 5. The worst case timing requirement of a task with checkpoints to tolerate k_i failures is minimized when its checkpoint intervals follow the scheme of Lemma 1 until $k_i - 1^{th}$ failure occurrence time, and then follow the scheme of Lemma 3.

Proof $\mathcal{A}_i(n, k_i)$ of a task τ_i can be formulated as follows:

$$\mathcal{A}_i(n, k_i) = n \cdot C_i + \sum_{j=1}^{k_i} (B_i(j) - S_i(j))$$

where $B_i(j)$ and $S_i(j)$ represent the increased and decreased timing requirement owing to j^{th} failure recovery, respectively. $S_i(j)$ is introduced because there is no more checkpointing overhead after the last permissible k_i^{th} failure occurs.

$$B_i(j) = \begin{cases} R_i + f_j, & \text{if } p_j = 0 \\ R_i + f_j - p_j - C_i, & \text{otherwise} \end{cases}$$
$$S_i(j) = \begin{cases} 0, & \text{if } 0 \leq j < k_i \\ (n - n_j) \cdot C_i, & \text{if } j = k_i \end{cases}$$

Thus,

$$\mathcal{A}_i(n, k_i) = R_i \cdot k_i + C_i \cdot (k_i + n_{k_i} - b)$$
$$+ \sum_{j=1}^{k_i} P_j \qquad (4)$$

Note that Eq. (4) takes the combined form of both Eq. (2) in the presence of $k_i - 1$ failures and Eq. (3). This fact shows that the scheme of Lemma 1 is able to minimize the worst case timing requirement of tasks until $k_i - 1^{th}$ failure occurrence time, and then the scheme of Lemma 3 is until the completion of tasks.

In this scheme, both the worst case timing requirement and the number of checkpoints are dependent on the number of checkpoints derived in the previous corollaries. These values vary according to nine different ranges which is determined by w_i, C_i and k_i. The detailed expression is skipped here because it can be calculated by simple algorithms based on the previous lemmas and corollaries. □

Until no more than $k_i - 1$ failures occur during the mission time, Theorem 5 scheme is equivalent to Lemma 1 scheme. But, the value of Eq. (4) is less than that of Eq. (2) owing to the reduced number of failure occurrences. After $k_i - 1^{th}$ failure occurrence time,

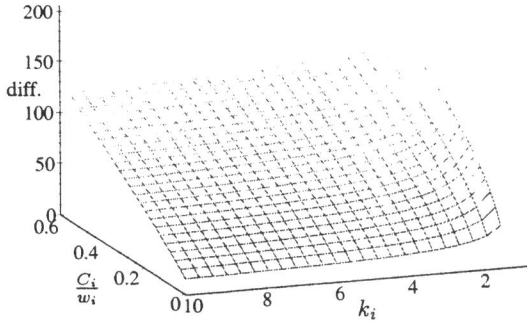

Fig. 1. $W_i(n^*, k_i)$ difference between equidistant scheme and Theorem 5 scheme

Theorem 5 scheme follows Lemma 3 scheme, which also makes the timing requirement less than Lemma 1 scheme. Fig. 1 illustrates that $W_i(n^*, k_i)$ difference between equidistant scheme and Theorem 5 scheme increases monotonously with C_i and the inverse of k_i where $w_i = 100$ and $R_i = 0.7 * C_i$. It shows that the gain is over 20% of w_i, so our scheme is advantageous in comparison with conventional checkpoint placement that is applicable to non real-time systems.

In order to obtain a good understanding of the effect of our scheme on timing requirement of tasks, a simulation has been conducted. $W_i(n^*, k_i)$ values of both our scheme and equidistant scheme are compared on two failure conditions; the first is that failures occur according to the worst case scenario, and the other is that failures occur randomly. All algorithms mentioned in this paper are implemented in the form of MAPLE scripts (Lee, 2000). The characteristics of tasks such as w_i, C_i and R_i have been generated randomly in the range $[100, 500]$, $[w_i/20, w_i/5]$ and $[C_i/2, C_i]$, respectively. For the fair comparison, equidistant scheme has been modified not to establish checkpoints after the last permissible failure occurs.

First, two thousand tasks have been generated for the comparison on the worst case failure condition. WCET of each task has been calculated according as the permissible number of faults varies from 1 to 5. Fig. 2 shows that, though $W_i(n^*, k_i)$ difference becomes smaller as the number of permissible failures increases, our scheme reduces WCET of task by about 7% to 26% compared with modified equidistant scheme. The reason that when $k_i = 1$ the difference is less than when $k_i = 2$ is that the gain caused by Lemma 1 scheme is added to that caused by Lemma 3 scheme.

Next, for the comparison of random failure condition, two thousand tasks have been also generated for each number of permissible failures. Upon calculation of the execution time of each instance, the average time is estimated. Fig. 3 shows its result that Theorem 5 scheme may reduce the average case performance.

The proposed checkpoint placement schemes can be implemented through timer interrupt routine of tasks or by incorporating a checkpointing server per

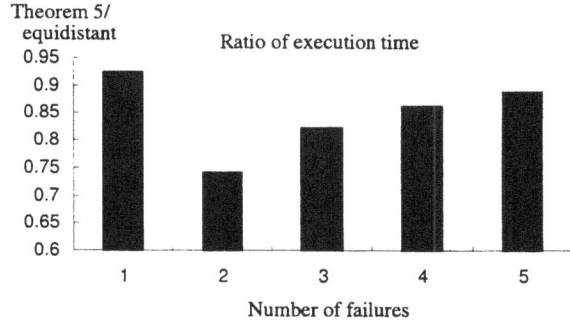

Fig. 2. $W_i(n^*, k_i)$ comparison on the worst case failure condition

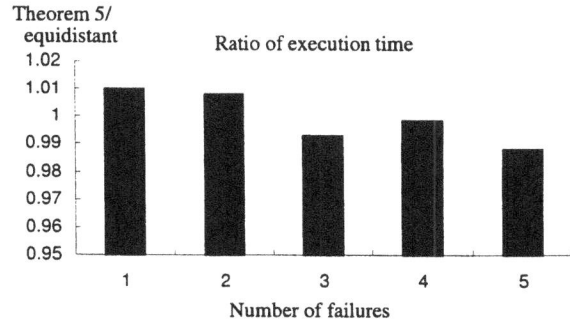

Fig. 3. $W_i(n^*, k_i)$ comparison on the random failure condition

task (Lee and Shin, 1999). The checkpointing server is released at every checkpoint interval time unit and its productive processing time equals C_i. Its priority should be higher than that of task under checkpointing and lower than that of more urgent tasks in order that all more urgent tasks be made to be completed before checkpointing is initiated.

5. CALCULATION OF THE PERMISSIBLE NUMBER OF FAULTS

The permissible number of faults k_i can be calculated from fault tolerance requirement. Commonly used fault tolerance requirement is expressed with minimum inter-arrival time T_F between two successive faults or the reliability goal of τ_i. This value for inter-arrival time between faults is either derived from past system fault data or assumed to be the worst case value the system can cope with. In case that a system designer can assume the minimum inter-arrival time T_F between two successive faults, k_i of τ_i with n checkpoints can be recursively derived from T_F as follows:

$$k_i(0) = \left\lceil \frac{w_i}{T_F} \right\rceil$$
$$k_i(j) = \left\lceil \frac{W_i(n, k_i(j-1))}{T_F} \right\rceil$$
$$k_i = \lim_{j \to \infty} k_i(j)$$

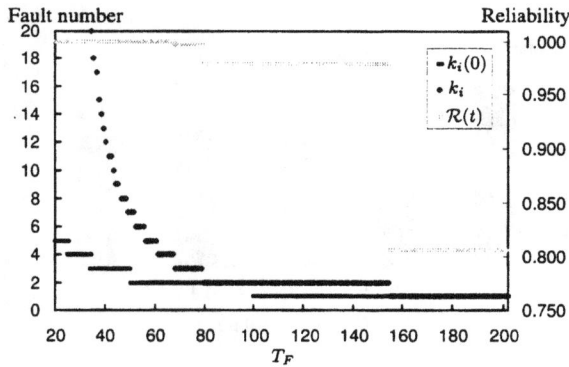

Fig. 4. Effect of T_F on k_i and $\mathcal{R}(t)$

T_F must be sufficiently large because, if not, k_i would converge into a large value. In case that a system designer wants a task τ_i to run successfully with more probability than a reliability goal ρ_i, k_i of τ_i can be derived as follows:

$$k_i = MIN\{j | \forall j \ \mathcal{R}(W_i(n,j)) > \rho_i\}$$

where the function $\mathcal{R}(t)$ calculates the reliability of a task. $\mathcal{R}(W_i(n,j))$ is equivalent to the probability for the task to suffer no more than j failures within $W_i(n,j)$. It can be obtained through k_i-stage Erlang distribution, expressed as

$$\mathcal{R}(W_i(n,j)) = \left(\sum_{i=0}^{j-1} \frac{(\lambda \cdot W_i(n,j))^i}{i!} \right) \cdot e^{-\lambda \cdot W_i(n,j)}$$

Fig. 4 illustrates an example how T_F is related with k_i and $\mathcal{R}(t)$ while $w_i = 100$, $C_i = 10$, $R_i = 7$, $n = n^*$ and $\lambda = 0.00139$. Other approaches usually use $k_i(0)$ as the maximum number of failures to occur during the mission time of the task. It is necessary, however, that the converging k_i be used because failure may occur while a task performs checkpointing or failure recovery.

6. CONCLUSION

This paper has quantified the worst case timing requirement of tasks that use checkpointing and derived the optimal placement for a task which has up to n checkpoints to tolerate the finite number of failures. The analysis shows that the equidistant scheme is not optimal checkpoint placement in real-time systems. Our simulation study shows that, at the expense of slight increase in the average case timing requirement of a task, our scheme reduces the worst case timing requirement of task by about 7% to 26% in comparison with the modified equidistant scheme. The timing information derived in this paper makes checkpointing applicable to real-time systems while keeping the validity of schedulability check developed in the past.

Our work is applicable to multiprocessors and distributed systems, if we assume that there exists an appropriate allocation scheme which statically allocates tasks to processors, and the scheduling is done on each processor independently using the uniprocessor schemes discussed in this paper.

7. REFERENCES

Bertossi, A.A. and L.V. Mancini (1994). Scheduling algorithms for fault-tolerance in hard real-time systems. *Journal of Real-Time Systems* 7(3), 229–245.

Bowen, N.S. and D.K. Pradhan (1993). Processor and memory based checkpoint and rollback recovery. *IEEE Computer* 26(2), 22–29.

Burns, A., R. Davis and S. Punnekkat (1996). Feasibility analysis of fault-tolerant real-time task sets. In: *Euromicro Workshop on Real-Time Systems*. pp. 29–33.

Castillo, X., S. McConnel and D. Siewiorek (1982). Derivation and calibration of a transient error reliability model. *IEEE Trans. Computers* 31(7), 658–671.

Ghosh, S. (1996). Guaranteeing Fault Tolerance Through Scheduling in Real-Time Systems. PhD thesis. University of Pittsburgh.

Iyer, R., D. Rossetti and M. Hsueh (1986). Measurement and modeling of computer reliability as affected by system activity. *ACM Trans. Computer Systems* 4(3), 214–237.

Lee, H. (2000). Scripts for checkpoint placement. http://cselab.snu.ac.kr/ffanta/main.html#ckptutil.

Lee, H. and H. Shin (1999). Worst case timing requirement of real-time tasks with time redundancy. In: *Proceedings of Real-Time Computing Systems and Applications*. pp. 410–413.

Long, J., W.K. Fuchs and J.A. Abraham (1990). A forward recovery strategy using checkpointing in parallel systems. In: *Proceedings of International Conference on Parallel Processing*. pp. 272–275.

Nicola, V.F. (1995). Checkpointing and the modeling of program execution time. In: *Software Fault Tolerance* (M.R. Lyu, Ed.). Chap. 7, pp. 167–188. John Wiley & Sons. Chichester.

Plank, J.S., M. Beck, G. Kingsley and K. Li (1995). Libckpt: Transparent checkpointing under unix. In: *Proceedings of USENIX Winter 1995 Technical Conference*. pp. 213–223.

Pradhan, D.K. and N. H. Vaidya (1994). Roll-forward checkpointing scheme: A novel fault-tolerant architecture. *IEEE Trans. Computers* 43(10), 1163–1174.

Punnekkat, S. (1997). Schedulability Analysis for Fault Tolerant Real-Time Systems. PhD thesis. University of York.

Shin, K.G., T. Lin and Y. Lee (1987). Optimal checkpointing of real-time tasks. *IEEE Trans. Computers* 36(11), 1328–1341.

Tantawi, A.N. and M. Ruschitzka (1984). Performance analysis of checkpointing strategies. *ACM Trans. Computer Systems* 2(2), 123–144.

Copyright © IFAC Distributed Computer Control Systems,
Sydney, Australia, 2000

DEADLINE ASSIGNMENT TO REDUCE OUTPUT JITTER OF REAL-TIME TASKS

Taewoong Kim, Heonshik Shin, and Naehyuck Chang

Computer Systems Laboratory
School of Computer Science and Engineering
Seoul National University, Seoul 151-742, Korea
{twkim,shinhs,naehyuck}@cse.snu.ac.kr
Voice: +82-2-874-1836, Fax: +82-2-874-3105

Abstract: In real-time control system, output jitter of a task is important when feedback control loops are implemented as execution of periodic tasks. Output jitter refers to the variation of response time of a periodic task. For the stability of the control system, output jitter should be reduced. This paper aims to reduce the output jitter of real-time tasks under EDF scheduling. The proposed scheme derives a linear program from execution times and periods of given tasks. To this end, this paper devises an objective function of linear program which reduces the output jitter and derives the constraints of linear program from the schedulability condition of EDF scheduling. The simulation measures the maximum and mean output jitter and relative output jitter to evaluate the performance of the proposed scheme. The simulation results show that the output jitter is reduced by 4-12% on average. *Copyright © 2000 IFAC*

Keywords: Real-time tasks, EDF scheduling, Deadlines, Jitter, Linear programming

1. INTRODUCTION

In real-time control systems, most of computational demands are periodic activities. For example, the periodic activities include data acquisition from the sensors, control loops, and system monitoring. In general, such periodic activities are implemented as execution of periodic tasks. Each periodic task τ_i is characterized by (c_i, p_i) where c_i and p_i denote execution time and period, respectively. A periodic task τ_i gives rise to an infinite sequences of jobs which arrive at time $(k-1) \cdot p_i$ where $k \geq 1$, respectively. In this case, a relative deadline (d_i) is assumed to be equal to p_i. The k-th job of τ_i arriving at time $(k-1) \cdot p_i$ should complete its execution until the time $k \cdot p_i$. The primary objective of real-time scheduling algorithm is to produce a feasible schedule such that periodic tasks complete prior to their respective deadlines. To this end, earliest deadline first (EDF) scheduling algorithm has been widely used in real-time systems (Liu and Layland, 1973). EDF scheduling algorithm dynamically assigns a higher priority to the task with earlier deadline and executes tasks in priority driven preemptive manner.

This paper aims to reduce the output jitter of real-time tasks under EDF scheduling. Reduction of output jitter can be the secondary objective of real-time scheduling algorithm. Output jitter refers to the variation of inter-completion times of consecutive jobs of the same task. In other words, output jitter represents the variation of response time of a task. The reduction of output jitter is important to control systems when feedback control loops are implemented with periodic tasks. In control theory, a feedback control loop assumes that feedback latency is constant (Shin and Kim, 1992; Ogata, 1995; Kim, 1998). In other words, there is no output jitter of periodic tasks. For the stability of the control system, output jitter should be reduced. This paper proposes a deadline assignment method which makes relative deadlines shorter than periods to reduce output jitter of periodic tasks. For example, a task τ_i can be executed isochronously if d_i is equal to c_i. But, such a deadline assignment may cause the other tasks to violates their deadlines. Hence, relative deadlines should be determined carefully to reduce the output jitter. This paper makes use of linear programming to determine the relative deadlines.

The rest of this paper is organized as follows: Section 2 reviews the related work. Section 3 defines the problem of deadline assignment, and devises a linear program to solve the problem. And then an illustrative example is shown. Section 4 evaluates the performance of the proposed algorithm. The paper concludes with Section 5.

2. RELATED WORK

Han and Lin have proposed a distance constrained task model where the distance represents an inter-completion time of two consecutive jobs of the same task (Han and Lin, 1992). Their approach applies a pinwheel problem to real-time task scheduling and transforms the periods of tasks into harmonic periods to bound the distance. Harmonic periods have the relation that p_i divides p_j when p_i is less than p_j (Kuo and Mok, 1991). However, the transformation of period increases the processor utilization. Hence, their scheme cannot be applied to a set of tasks with high processor utilization (Lin and Herkert, 1996).

Under EDF scheduling algorithm, Mauthe and Coulson have dealt with the scheduling of jitter constrained threads for multimedia applications and have analyzed the schedulability of threads (Mauthe and Coulson, 1997). In their paper, the deadlines of threads are earlier than their periods to reduce the jitter. But, the problem of deadline assignment has not been addressed. Schedulability analysis of tasks whose deadlines are shorter than their periods has also been analyzed in (Baruah et al., 1990; Ripoll et al., 1996). This paper exploits the schedulability analysis in (Ripoll et al., 1996).

Baruah et al. have addressed the problem of deadline assignment to reduce the output jitter of tasks (Baruah et al., 1999). Their scheme can reduce the output jitter for small number of tasks. The aim of this paper is to reduce the output jitter of whole tasks in a given set.

3. DEADLINE ASSIGNMENT

It is assumed that there is a set of periodic tasks $T = \{\tau_1, \tau_2, \ldots, \tau_n\}$. Each task τ_i in T is characterized by (c_i, p_i). The problem of output jitter reduction can be described as follows:

Problem: *Finds feasible relative deadlines* $(d_1, d_2, \ldots, d_n)$ *of tasks in* T *where* $c_i \leq d_i \leq p_i, 1 \leq i \leq n$ *to have small output jitter.*

Since $d_i \leq p_i$, a task τ_i can meet its original timing requirement. This problem cannot be solved by an enumerative method which permutes all possible relative

deadlines and checks the schedulability iteratively. Such an enumerative method requires $\prod_{i=1}^{n}(p_i - c_i)$ iterations, hence, it cannot find a solution in a reasonable time.

To effectively derive the relative deadlines, this paper exploits an integer linear programming technique. A linear program (LP) is an optimization problem with a linear objective function and a set of linear constraints as follows (Ahuja et al., 1993):

Minimize $\quad Z = C_1x_1 + C_2x_2 + \cdots + C_nx_n$
subject to $\quad a_{11}x_1 + a_{12}x_2 + \cdots + a_{1n}x_n \leq b_1$
$\qquad\qquad a_{21}x_1 + a_{22}x_2 + \cdots + a_{2n}x_n \leq b_2$
$$\vdots$$
$\qquad a_{m1}x_1 + a_{m2}x_2 + \cdots + a_{mn}x_n \leq b_m$
$\qquad\qquad \forall i, lb_i \leq x_i \leq ub_i$

where $(x_1, x_2, \ldots, x_n)$ is the vector of variables to be solved, Z is called an objective function, the inequalities are called the constraints, and C_i, a_{ij}, b_i, lb_i, and ub_i are known constants, respectively. An integer linear program refers to a linear program whose variables are constrained to integers. Fig. 1 shows the outline of the solution approach. To formulate an integer linear program with variables $(d_1, d_2, \ldots, d_n)$, an linear objective function and linear constraints should be derived from execution times and periods.

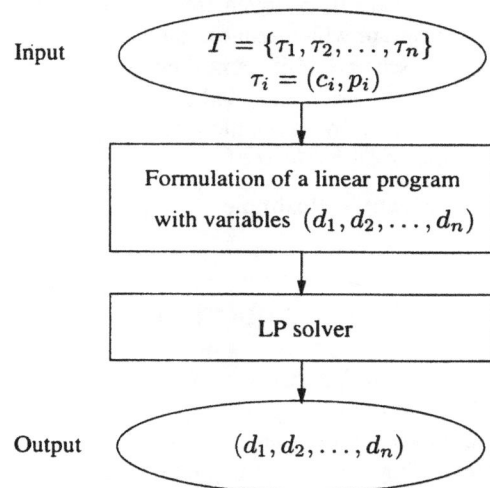

Fig. 1. Overview of solution approach

3.1 *Derivation of objective function*

Before developing an objective function to reduce the output jitter of tasks, output jitter of a task can be quantified as follows:

- Maximum output jitter
- Relative output jitter

The maximum output jitter represents the maximum variation of inter-completion time of a task from its period. As shown in Fig. 2, the maximum inter-completion time (IC_i^{max}) can amount to $p_i - c_i +$

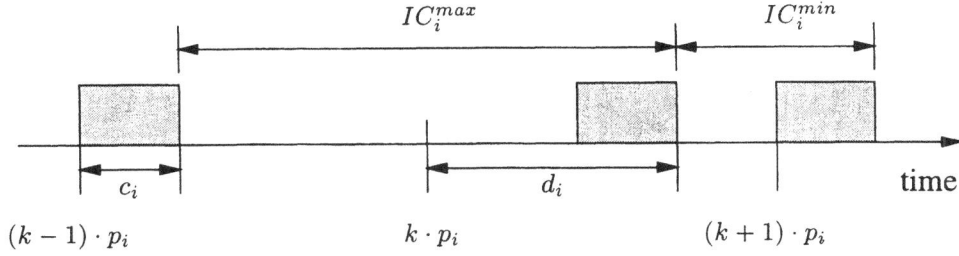

Fig. 2. The output jitter of τ_i

d_i and the minimum inter-completion time (IC_i^{min}) can amount to $p_i - d_i + c_i$, respectively. Thus, the maximum output jitter of τ_i, denoted by J_i, can be written in Eq. (1).

$$J_i = \max(IC_i^{max} - p_i, p_i - IC_i^{min})$$
$$= d_i - c_i \quad (1)$$

Based on the maximum output jitter, the relative output jitter of τ_i is defined as Eq. (2).

$$\text{relative output jitter} = \frac{J_i}{p_i} = \frac{d_i - c_i}{p_i} \quad (2)$$

Since the goal of this paper is to reduce the output jitter of whole tasks in T, the objective function of a linear program should be defined to minimize the sum of relative output jitters.

$$Z = \sum_{i=1}^{n} \frac{d_i - c_i}{p_i} = \sum_{i=1}^{n} \frac{d_i}{p_i} - \sum_{i=1}^{n} \frac{c_i}{p_i} \quad (3)$$

In Eq. (3), the second term $(\sum_{i=1}^{n} c_i/p_i)$ is a constant. Therefore, the objective function Z is rewritten as follows:

$$Z = \sum_{i=1}^{n} \frac{d_i}{p_i}. \quad (4)$$

3.2 Derivation of linear constraints

This section describes linear constraints of relative deadlines for linear programming. The constraints of linear program should be derived from the schedulability condition of EDF scheduling. When the relative deadline is equal to period, the schedulability test of periodic tasks is simple. It is shown that EDF scheduling algorithm always guarantees the deadlines of periodic task unless the processor is overloaded. However, the schedulability test has a pseudo-polynomial time complexity when the relative deadline is shorter than period (Baruah et al., 1990; Ripoll et al., 1996). In (Ripoll et al., 1996), $H(t)$ denotes the amount of execution time that has been requested by all tasks whose deadlines are less than or equal to t.

$$H(t) = \sum_{j=1}^{n} \left\lfloor \frac{t + p_j - d_j}{p_j} \right\rfloor \cdot c_j \quad (5)$$

The schedulability condition is that the periodic tasks can always meet their deadlines if the following inequalities are satisfied (Ripoll et al., 1996).

$$\forall i, 1 \leq i \leq n, H(d_i) \leq d_i \quad (6)$$

However, the inequalities in Eq. (6) cannot be applied to a linear program, because they have nonlinear floor functions. These inequalities can have linear constraints by approximating the floor function. Since $\lfloor \frac{d_i + p_j - d_j}{p_j} \rfloor = 0$ if $d_i < d_j$, $H(d_i)$ can be rewritten as

$$H(d_i) = \sum_{j=1}^{n} \left\lfloor \frac{d_i + p_j - d_j}{p_j} \right\rfloor \cdot c_j$$
$$= \sum_{\forall j, d_j \leq d_i} \left\lfloor \frac{d_i + p_j - d_j}{p_j} \right\rfloor \cdot c_j.$$

Since $\lfloor x \rfloor \leq x$, the inequalities given in Eq. (6) are approximated as follows:

$$\forall i, \text{ where } 1 \leq i \leq n,$$
$$H'(d_i) = \sum_{\forall j, d_j \leq d_i} \left(\frac{d_i + p_j - d_j}{p_j} \right) \cdot c_j \leq d_i. \quad (7)$$

Although the inequalities in Eq. (7) are linear with respect to variables $(d_1, d_2, \ldots, d_n)$, such an approximation cannot be made without knowledge of relative deadlines. To resolve this problem, the proposed scheme enforces the following relations on the relative deadlines when the constraints are formulated.

$$\forall i, j, \ d_i \leq d_j \text{ iff } c_i \leq c_j$$

From the above discussion, the constraints of linear program can be written as given in Eq. (8).

$$\forall i, \text{ where } 1 \leq i \leq n,$$
$$\sum_{\forall j, c_j \leq c_i} \left(\frac{d_i + p_j - d_j}{p_j} \right) \cdot c_j \leq d_i \quad (8)$$

By definition, the lower and upper bounds on d_i are given in Eq. (9).

$$\forall i, 1 \leq i \leq n, c_i \leq d_i \leq p_i \quad (9)$$

Thus, all linear constraints of the linear program have been formulated in Eqs. (8) and (9).

Table 1. Example tasks

Task	Execution time	Period
τ_1	1	3
τ_2	3	5

Table 2. Reduction of output jitter

Performance Measure	EDF with $d_i = p_i$		Proposed scheme	
	τ_1	τ_2	τ_1	τ_2
Maximum output jitter	1.0 time unit	1.0 time unit	1.0 time unit	0.0 time unit
Relative output jitter	0.33	0.20	0.33	0.0
Mean output jitter	0.11 time unit	0.2 time unit	0.0 time unit	0.0 time unit

3.3 *Illustrative example*

This section shows a simple example of derivation of relative deadlines. Consider a set of tasks shown in Table 1. According to Section 3.1 and 3.2, the following integer linear program can be formulated.

$$\begin{aligned} \text{Minimize} \quad & Z = \tfrac{1}{3}d_1 + \tfrac{1}{5}d_2 \\ \text{subject to} \quad & -d_1 \le -1 \\ & -\tfrac{1}{3}d_1 - \tfrac{2}{3}d_2 \le -4 \\ & 1 \le d_1 \le 3 \\ & 3 \le d_2 \le 5 \end{aligned}$$

The objective function is obtained from Eq. (4) and the constraints are from Eqs.(8) and (9), respectively. The above integer linear program is solved by a free software called lp_solve (Berkelaar, 1998). The lp_solve accepts the following input format to derive the relative deadlines.

```
min: 0.333 d1 + 0.2 d2;
 -1.0 d1 <= -1.0;
 -0.333 d1 -0.667 d2 <= -4.0;
 d1 >= 1.0;
 d2 >= 3.0;
 d1 <= 3.0;
 d2 <= 5.0;
 int d1, d2;
```

Finally, the lp_solve produces the result as follows:

Solution: $(d_1, d_2) = (2, 5)$.

Table 2 shows the quantities related with output jitter of tasks when the relative deadlines are equal to periods and are assigned by the proposed scheme, respectively. According to Table 2, the proposed scheme reduces the maximum output jitter and relative output jitter of tasks compared to EDF scheduling with $d_i = p_i$. The proposed scheme also reduces the mean output jitter. Since the mean output jitter is equal to 0, the mean inter-completion time of a task is equal to its period.

4. PERFORMANCE EVALUATION

4.1 *Experimental setup*

This section evaluates the performance of the proposed scheme through simulation for various task sets. For the simulation, the following parameters are used to generate task sets randomly.

- The number of task sets: 100
- The number of tasks in a set: 5 - 10
- The period of a task: 10 - 250 (multiples of 10)
- Processor utilization: 50%, 70%, and 90%

The execution times of tasks are adjusted according to processor utilization. In the simulation, the three following performance measures have been measured.

- Maximum output jitter
- Relative output jitter
- Mean output jitter

The simulation has also measured the performance of distance constrained task model (Han and Lin, 1992).

4.2 *Experimental results*

First, Fig. 3 shows the reduction of the maximum output jitter compared to EDF scheduling. Note that the values shown in the graphs represent the mean values. Fig. 3 shows that the proposed scheme reduce the maximum output jitter by up to 18.5% (with an average of 12.9%).

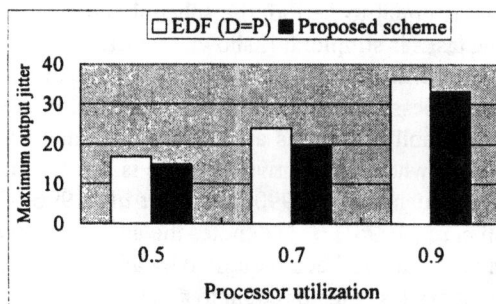

Fig. 3. Maximum output jitter

Table 3. Performance of distance constrained task scheduling

Processor utilization	Mean increase of utilization	Number of transformed task sets
50 %	16.68 %	100
70 %	21.21 %	100
90 %	N/A	0

Second, Fig. 4 shows the comparison of relative output jitter. The performance gain obtained from the proposed scheme is relatively small, since periods are much greater than output jitter. The relative output jitter can be reduced by up to 7.2% (with an average of 4.1%).

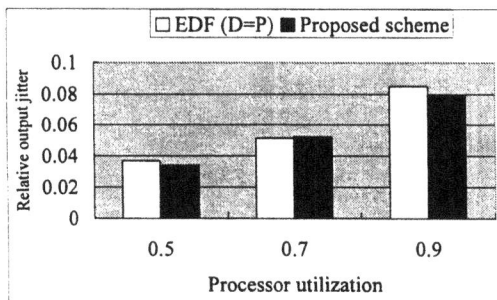

Fig. 4. Relative output jitter

Third, Fig. 5 shows the improvement of the mean output jitter by the proposed scheme. Fig. 5 shows that the proposed scheme reduces the mean output jitter by up to 15.2% (with an average of 10.6%). In summary, the proposed scheme reduces the output jitter of tasks by 4-12% on average.

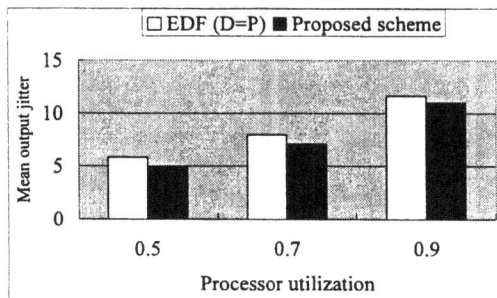

Fig. 5. Mean output jitter

Table 3 shows the performance of distance constrained task scheduling (DCTS). If a set of tasks is transformed by DCTS, the transformed tasks produce no output jitter. But, the transformed tasks are executed with different periods from the original. The experiment measures the number of task sets which can be transformed by DCTS and the increase of processor utilization, respectively. In cases of 50% and 70% processor utilization, DCTS can schedule 100 task sets successfully. However, the processor utilization increases by 18.9% on average. Thus, DCTS cannot

schedule any task set when the processor utilization is 90%. According to Table 3, the proposed scheme is superior to DCTS when tasks have high processor utilization.

5. CONCLUSION

This paper proposes a deadline assignment scheme which adjusts relative deadlines to reduce the output jitter of periodic real-time tasks under EDF scheduling. The proposed scheme derives an integer linear program from execution times and periods of given tasks to effectively find a feasible relative deadline vector. To this end, this paper proposes an objective function of linear program which reduces the output jitter and derives the constraints of linear program from the schedulability condition of EDF when $d_i \leq p_i$. From experimental results in Section 4, it is shown that the proposed assignment scheme reduces the output jitter compared to EDF scheduling with $d_i = p_i$.

6. REFERENCES

Ahuja, R. K., T. L. Magnanti and J. B. Orlin (1993). *Network flows Theory, Algorithms, and Applications*. Prentice-Hall.

Baruah, S. K., G. Buttazzo, S. Gorinsky and G. Lipari (1999). Scheduling periodic task systems to minimize output jitter. In: *Proceedings of the Real-Time Computing Systems and Applications*. pp. 62–69.

Baruah, S. K., R. Howell and L. E. Rosier (1990). Algorithms and complexity concerning the preemptive scheduling of periodic, real-time tasks on one processor. *Real-Time Systems* **2**, 301–324.

Berkelaar, Michel (1998). lp_solve version 2.3. ftp://ftp.ics.ele.tue.nl/pub/lp_solve.

Han, C-C. and K-J. Lin (1992). Scheduling distance-constrained real-time tasks. In: *Proceedings of the Real-Time Systems Symposium*. pp. 300–308.

Kim, Byung Kook (1998). Task scheduling with feedback latency for real-time control systems. In: *Proceedings of the Real-Time Computing Systems and Applications*. pp. 37–41.

Kuo, Tei-Wei and A. K. Mok (1991). Load adjustment in adaptive real-time systems. In: *Proceedings of the Real-Time Systems Symposium*. pp. 160–170.

Lin, K-J. and A. Herkert (1996). Jitter control in time-triggered systems. In: *Proceedings of Hawaii International Conference on System Sciences*.

Liu, C. L. and J. W. Layland (1973). Scheduling algorithms for multiprogramming in a hard real-time environment. *Journal of the ACM* **20**(1), 46–61.

Mauthe, A. and G. Coulson (1997). Scheduling and admission testing for jitter-constrained periodic threads. *Multimedia Systems* **5**(5), 337–346.

Ogata, Katsuhiko (1995). *Discrete-Time Control Systems*. 2nd ed.. Prentice-Hall International, Inc.

Ripoll, I., A. Crespo and A. K. Mok (1996). Improvement in feasibility testing for real-time tasks. *Real-Time Systems* **11**(1), 19–39.

Shin, Kang G. and Hagbae Kim (1992). Derivation and application of hard deadlines for real-time control systems. *IEEE Transactions on Systems, MAN, and Cybernetics* **22**(6), 1403–1413.

Copyright © IFAC Distributed Computer Control Systems,
Sydney, Australia, 2000

CLOCK SYNCHRONIZATION PROTOCOL UNDER WINDOWS NT

Armando D. Assandri (*) and Miguel A. García (**)

(*) Instituto de Automática – Universidad Nacional de San Juan
Av. Libertador 1109 Oeste - J5400ARL San Juan – Argentina
aassandr@inaut.unsj.edu.ar
(**) Dpt. of Systems Engineering and Automatic Control - University of Valladolid
Prado de la Magdalena s/n – 47005 Valladolid – Spain
miguel@autom.uva.es

Abstract: Synchronization problems are a well-known key topic in Distributed Control Systems. Most of the functionalities are implicitly based on the time consistency along the system. At the same time, general purpose operating systems, and Windows NT paradigmatically, have appeared in the control scene with undeniable strength. In fact, most of the SCADAs and industrial DCSs have migrated to this kind of platform. In this paper a clock synchronization protocol at driver level, based on the master/slave paradigm and using a dedicated and determinist channel, is presented and quantitative results are summarised. The time discrepancy results are applied to the clock correction at the slave by using a control algorithm. The given solution is based on a standard peripheral and so it can be considered as a benchmark result. *Copyright © 2000 IFAC*

Keywords: Clock sinchronization; Drivers; Protocols; Distributed control, Operating Systems.

1. INTRODUCTION

During the last years, there is a growing interest in the industry to adopt Windows NT as a front end environment in a wide variety of control applications. Even though it can't be considered as a hard real time operating system, some effort is being dedicated to overcome its drawbacks. New products like *HyperKernel*, *INtime* and *RTX* give a step forward in this sense. Other works like Ramamritham et al. (1998) demonstrate that Windows NT can be used as a *soft real time system*, perfectly suitable for most low dynamics control processes.

Besides, Windows NT gives support to Remote Procedure Calls (RPCs), owns embedded networking capabilities and provides all the basic tools needed to develop distributed computer control applications. Most of DCSs algorithms are based on time consistency along the network, and therefore time

sinchronization is a key aspect to be considered in order to obtain good results.

Existing protocols like NTP (Network Time Protocol) (Mills, 1991) provides enough precision for a range of applications. According with published marks of this protocol, it can reach a precision in the sinchronization in the order of tenths of milliseconds. The main drawback of the NTP protocol, is that it is based on the TCP/IP protocol. The sinchronization is made using statistical values of the delay existing in the channel that connects the synchronized computer with the reference computer.

The present work was developed trying to reach a precision in the order of a few microseconds, for those higher dynamics applications that demands a tighter adjustment in the system time. In order to obtain a very low delay in the communication channel, two drivers were used to send timestamps from the reference computer to the sinchronized computer. A user level process is then used to maintain the error within a bounded margin.

2. THE LOW LEVEL PROTOCOL

A deterministic approach was used to carry out the low level protocol (at the physical layer) in order to avoid the uncertainties introduced by network delays. As the parallel port is a common and simple peripheral found in any PC type computer, it was used to implement the connection between two computers to be synchronized.

Among the several modes of operation and connection possibilities for this port, the bidirectional ECP byte mode was chosen. The communication is interrupt driven. A periodic task takes a 64 bits timestamp in the reference computer and then split it in 8 bytes. Afterwards, the bytes are transmitted one by one to the synchronized computer by using interrupts.

The protocol uses 3 lines for the transmission control and 2 lines for interruptions. Besides, two output lines are used to trigger an universal counter in order to measure the transmission delay. The channel is unidirectional, dedicated, with low traffic and without media access collisions. Table 1 shows a summary of the lines used and its purpose and Fig. 1

shows a timing diagram of the protocol at the physical layer.

3. METHODOLOGY

The methodology proposed is based on the work developed by Gergeleit and Streich (1994) and is depicted in Fig. 2. The reference computer takes reference timestamps periodically. The sampling period can be changed from a User level process. The reference timestamp is transmitted through the parallel port to the slave driver. When the first byte reaches the synchronized computer, the slave driver takes the local timestamp. After the last byte arrives to the slave driver, the error between the clock of the two computers is computed. The control process that runs at User level waits for an event that is signaled after the error is calculated. Then the process reads the error with an I/O Request Packet (IRP) and the control algorithm is applied in order to reduce the error to zero.

An universal counter is used to measure the time delay of the transmission. These values can be added to the error in the control process to compensate both the time used in the transmission and the response time of the slave Interrupt Service Routine (ISR). The values are deterministic, but some small deviations may appear due to the random influence of other ISRs belonging to device drivers with higher priorities.

Table 1 Parallel port connections.

Master	Pin #	Direction	Pin #	Slave	Function
D0 ~ D7	2 ~ 9	⇒	2 ~ 9	D0 ~ D7	Data bits.
C0-	1	⇒			Trigger edge for delay measure.
C1-	14	⇒	12	S5+	Active transmission line (frame).
C2+	16	⇒	10	S6+	Master to slave interrupt line.
C3+	17	⇒	11	S7-	Byte 0 line.
		⇐	1	C0-	Trigger edge for delay measure.
S6+	10	⇐	16	C2+	Slave to master interrupt line.
S7-	11	⇐	17	C3-	Slave ready line.

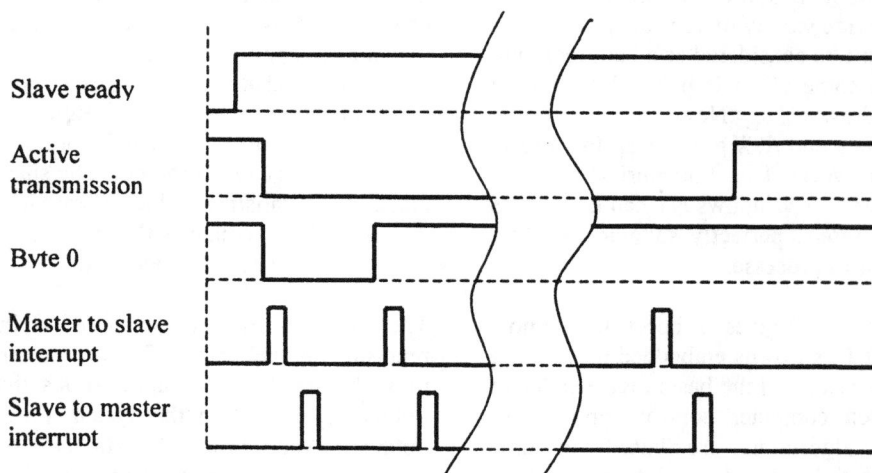

Fig. 1 Timing diagram of the physical layer.

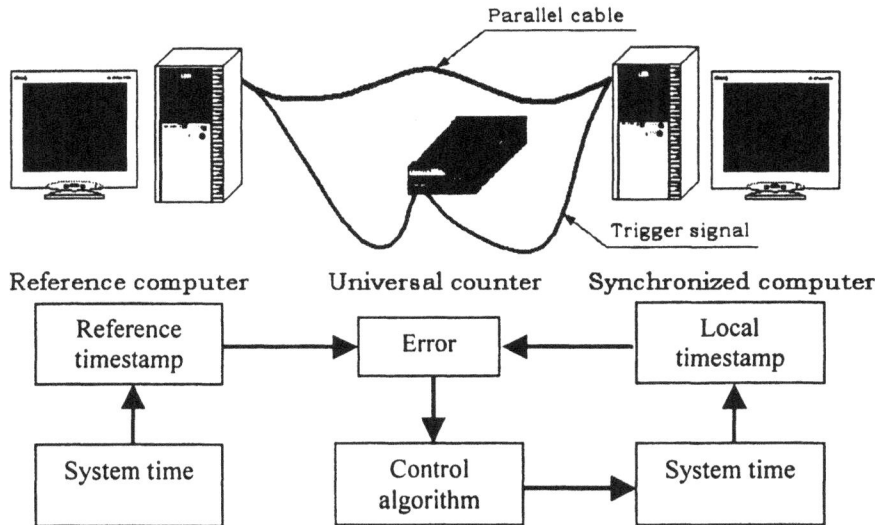

Fig. 2 Methodology used to synchronize the clock.

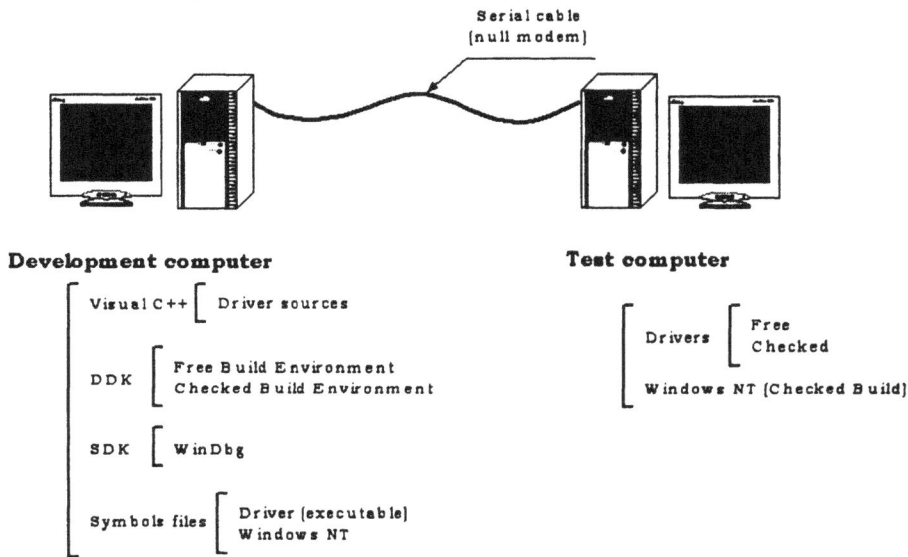

Fig. 3 Development system.

4. THE DRIVERS

It was necessary to use two kernel mode drivers to control the whole transaction through the parallel port in order to achieve the minimum transmission delay. The implementation follows the master/slave paradigm, being the master driver in the reference computer and the slave driver(s) in the synchronized computer(s). A development system built with all the Microsoft's standard tools (SDK, DDK and Visual C++) was used to implement and test the drivers. Fig. 3 shows such a system and all the tools and Windows NT versions used during the debugging operations.

Within each driver, several kernel objects are created in order to achieve a proper functionality. Besides the device object associated with the parallel port, we can mention: the Interrupt object used to connect the

ISR with an interrupt vector, the DPC (Deferred Procedure Call) object and the Event object. The driver's structure follows the general guidelines given in Viscarola and Mason (1999) and in Dekker and Newcommer (1999).

User mode application programs can interchange data with the driver through the I/O Manager. The standard procedure uses IRPs to read/write data from/to the driver. Both drivers have proper dispatch routines to manage I/O requests. In Fig. 4 it is shown the way the drivers interact between them through the parallel port protocol and with User mode applications.

Both drivers have a DPC and an ISR. In Fig. 5 there is a timing diagram that shows how these routines interact during the transmission of a timestamp.

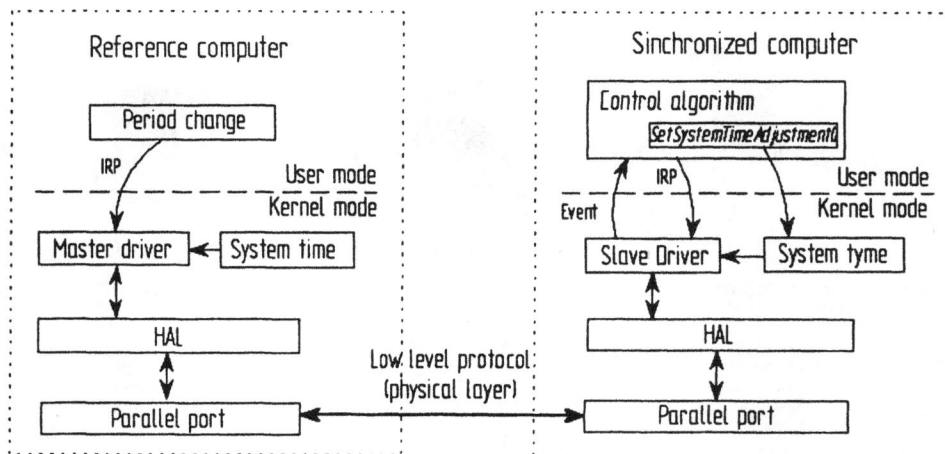

Fig. 4 Scheme of the synchronization process.

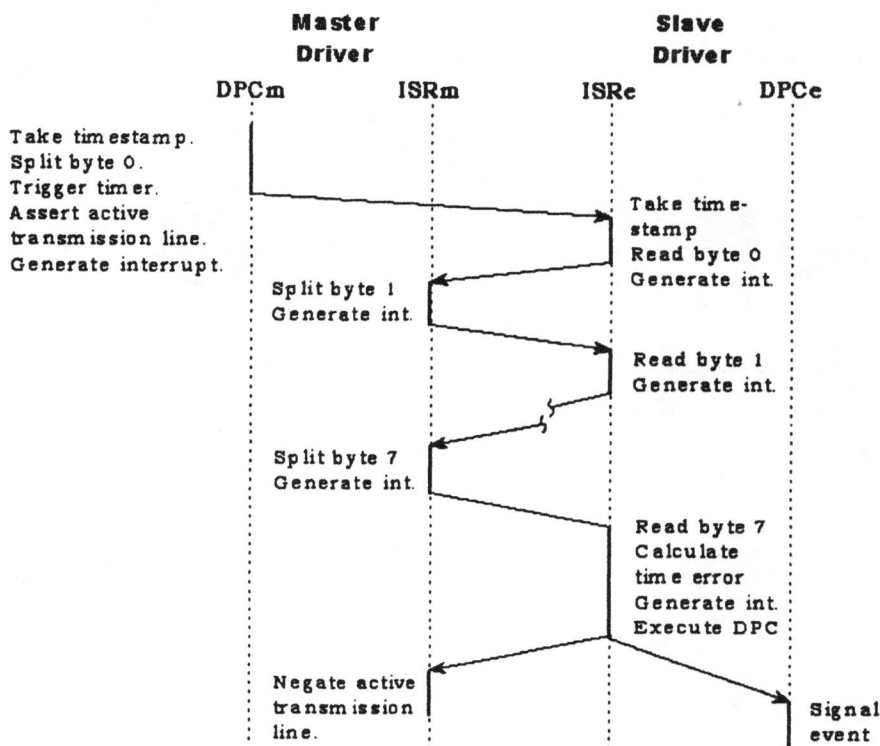

Fig. 5 Timing diagram of the DPCs and the IRPs.

5. WINDOWS NT SYSTEM TIME

The system clock in Windows NT is the result of a combination of hardware and software. A hardware timer generates interruptions periodically (which is known as a *clock tic*). The period between interruptions is 10 ms in the version for one processor and 15 ms in the multiprocessor version (only in x86 microprocessor family based computers). The register that holds the system time is upgraded with each interruption by adding a fixed amount (100,144 in the 10 ms version), provided that no correction is being applied to the clock. This register is a 64 bit signed enter and the unit represents 100 ns. The system clock uses the UTC time scale with the origin (the epoch corresponding to the zero value) at the 01/01/1601.

6. CONTROL ALGORITHM

The DDK (Device Driver Kit) provides no function to set the system time from a kernel mode driver. At this level the only way to get a timestamp is the function *KeQuerySystemTime()*. This is not the case with user mode applications (Win32 API), where we can find two functions to change the system time, provided that the process have the appropriate rights to do so. These functions are: *SetSystemTime()* and *SetSystemTimeAdjustment()*. The first one sets the time with a value passed as a parameter to the function. The second one allows small changes in the system time in order to synchronize it with a reference clock. This function has two parameters: one to enable or disable the time adjustment and the other to set the amount to add to the system clock register with each tic.

Based on these functions, two different control strategies are used depending on the order of the error. If the error is bigger than ±60 seconds, it is added to the current time and then the resulting value is passed to the function *SetSystemTime()* as a parameter. The time is corrected in just one step, and then the other part of the control algorithm is applied.

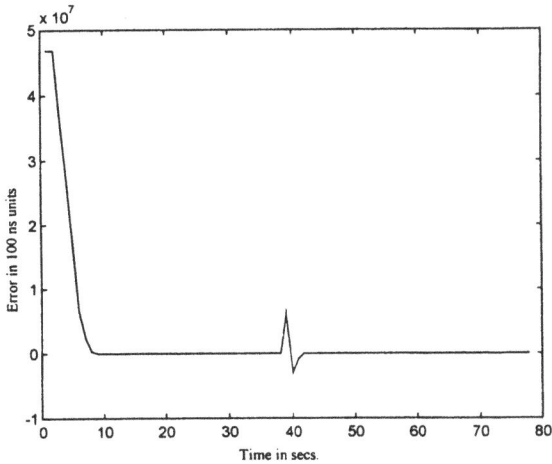

Fig. 6 Evolution of the error with disturbances.

If the error is less or equal than ±60 seconds, a gradual strategy is applied using the function *SetSystemTimeAdjustment()*. In this case, the control algorithm works with the same policy of a finite time controller algorithm: it tries to reduce the error to zero in a finite amount of time, mainly in just one sampling period. The error is distributed proportionally in all the tics that are included in one sampling period. As this strategy could produce very high correction values when the clock is too shifted, a saturation function is applied if the correction to be

applied is bigger than ± 50% of the normal value used to upgrade the system clock. So, the saturation function only allows values in the band between 50,072 and 150,216 for the first parameter of the *SetSystemTimeAdjustment()* function. Fig. 6 shows the behaviour of this strategy in the presence of imposed disturbances.

The gradual algorithm can reduce the error to the order of a few microseconds. It has a steady state error that is a linear function of the sampling period used to take the timestamps. With a sampling period of 1 sec. the steady state error is bounded to 10 μs.

7. THE GRANULARITY PROBLEM

Kopetz and Ochsenreiter (1987) defined the granularity of the clock as the period of its oscillation. As mentioned before, in Windows NT the system clock is upgraded with a period of 10 (or 15) ms, and the time remains constant between clock tics. Therefore, the time function is non linear and resembles a stair shape

The circuit that generates the interruptions used to upgrade the time starts when the computer is powered on. So, even though two different computers were turned on at the same time, the probability of upgrading its time at different moments is high, and there is a phase shift between their clocks. Almost all the methods found in the literature to synchronize the clock are based on timestamps that normally do not convey phase information. Thus, when only time information is used to synchronize the clock, the phase shift implies an error on its own. Fig. 7 shows a possible scenario before and after the synchronization is carried out. It can be clearly seen that even though the steady state error is in the order of a few μs, there is a much bigger error due to the

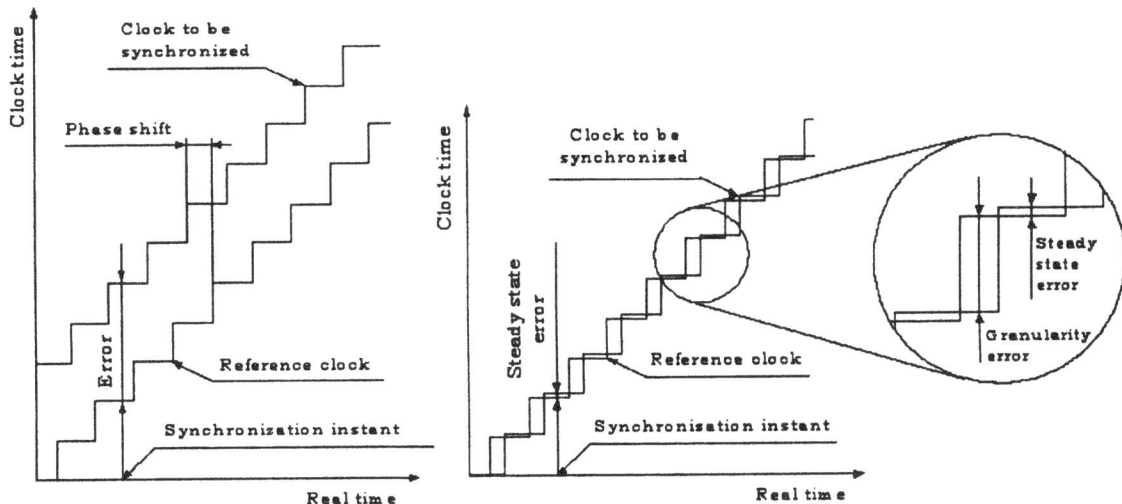

Fig. 7 The system clock before and after the synchronization.

granularity (in the order of the ms). This error is bounded by the granularity of the system clock.

8. TESTS AND RESULTS

The protocol was tested in several computers under different charge conditions. Five charge types were chosen to take measures of the transmission delay with the universal counter:

1) No charge.
2) Hard disk check.
3) File download from Internet.
4) A multimedia application from Internet.
5) A Matlab simulation.

Some figures of the transmission delay are summarised in Table 2. The universal counter allows statistical calculations of the values acquired during the experiment. The statistics shown are based on 100 samples.

Table 2. Transmission delay

Charge type	1	2	3	4	5
Average μs	25,0	26,1	26,9	27,1	25,6
Maximum μs	25,9	28,9	28,7	34,6	26,8
Minimum μs	23,9	24,7	25,1	24,5	24,8
σ		0,32	0,73	0,67	1,04	0,35

The values shown in Table 2 represent the time elapsed since the master driver takes the reference timestamp until the slave driver takes the local timestamp. The protocol also allows the measure of the time used to transmit one timestamp. These values have an average of 220 μs.

9. CONCLUSIONS

The developed protocol uses a combination of two Kernel mode drivers for the parallel port and a control algorithm running in User mode. The steady state error achieved after less than two minutes is bounded by a linear function of the timestamps sampling rate. With sampling rates of a few seconds, the steady state error is under 0,5% of the error due to the granularity of the system clock.

The main conclusion of this work is that it is not only important to achieve a high speed transfer rate for the timestamps and a known and deterministic transmission delay in the communication channel. The granularity of the system clock bounds the precision that can be reached when two computers are synchronized under Windows NT.

REFERENCES

Dekker E. N., Newcomer J. M. (1999) *Developing Windows NT Device Drivers*. Addison-Wesley, Massachusetts.

Gergeleit M., Streich H. (1994). Implementing a Distributed High-Resolution Real-Time Clock Using the CAN-Bus. *1st International CAN Conference*, Erlangen.

Kopetz H., Ochsenreiter W. (1987) Clock Synchronization in Distributed Real-Time Systems. *IEEE Tran. on Computers*, **C-36 8**, 933-939.

Mills, D. L. (1991). On the chronometry and metrology of computer network timescales and their application to the Network Time Protocol. *ACM Computer Communications Review*, **21,5**, 8-17.

Ramamritham, K. et al. (1998) Using Windows NT for Real-Time Applications: Experimental Observations and recommendations. *Proceedings of the Fourth IEEE Real-Time Technology and Applications*, Denver, CO.

Viscarola P. G, Mason W. A. (1999) *Windows NT Device Driver Development*. MacMillan Technical Publishing, Indianapolis.

Copyright © IFAC Distributed Computer Control Systems,
Sydney, Australia, 2000

INTEGRATING WCET ANALYSIS INTO A MATLAB/SIMULINK SIMULATION MODEL [1,2,3]

Raimund Kirner* Roland Lang** Peter Puschner* Christopher Temple*

Technische Universität Wien, Austria
***Dependable Computer Systems KEG, Wien, Austria*

Abstract:
Traditional worst-case execution time (WCET) analysis interfaces to the user either
through high-level language source code or assembly/machine code. This paper
demonstrates how WCET analysis can be integrated into high-level application
design and simulation tools like Matlab/Simulink, thus providing a higher-level
interface to WCET analysis. The paper shows necessary restrictions and adaptions to
Matlab/Simulink that make code generated from the tool chain amenable to WCET
analysis. It presents the interface between the Matlab/Simulink block set and WCET
analysis and explains the steps performed during WCET analysis.
Copyright ©2000 IFAC

Keywords: Worst-Case Execution Time Analysis, Execution Times, Performance
Analysis, Real-Time Languages, Real-Time Computer Systems, Simulation, Model
Management

1. INTRODUCTION

The knowledge of the worst-case execution time
(WCET) of programs is crucial for the design of
hard real-time systems. Therefore, WCET analysis has become an acknowledged part in the design
of real-time systems in the last years.

The first WCET analysis concepts were based on
the analysis of object/assembly code (Zhang *et
al.*, 1993; Harmon *et al.*, 1994; Li *et al.*, 1995). Additional information required for WCET analysis
was given by annotations inside the assembly code

or interactive user input. This was an acceptable
approach for the analysis of embedded systems
which were commonly written in an assembly language.

As hardware became cheaper, also the developing
environments had been improved. Applications
became more often to be written in a high-level
programming language. WCET analysis methods
have been adopted to provide additional timing
information by the programmer at the high-level
soure code. Actually, there exist WCET interfaces
for high-level languages like Euclid (Klingerman
and Stoyenko, 1986), Modula2 (Vrchoticky, 1994),
ADA (Bernat *et al.*, 2000), C (Park, 1993; Kirner,
2000), etc.

At present a very novel topic of software development is the model-based specification from which
code generation can be performed automatically.
WCET analysis can be integrated into such a
framework with extended capabilities by taking
advantage of the structure of the models. An example for this can be found in (Erpenbach and

---

[1] This work has been supported by the IST research
project "Systems Engineering for Time-Triggered Architectures (SETTA)" under contract IST-10043.
[2] Copyright 2000 IFAC. This electronic preprint made
available by the author as a courtesy. For further publication rights contact IFAC or the author. This article
was presented at the 16th IFAC Workshop on Distributed
Computer Control Systems, Sydney, Australia, November,
2000.
[3] MATLAB and Simulink are registered trademarks of the
Mathworks, Inc.

Altenbernd, 1999) for the Statemate statechart system (Harel and Naamad, 1996).

The work in this paper presents an approach of integrating WCET analysis into a model simulation system with capabilities of code generation for rapid prototyping. The general WCET analysis approach is based on introducing timing information at a *high representation-level* of a system and performing the WCET analysis at *assembly-language level*.

The WCET analysis discussed in this work is targeted to determine the runtime behavior of a Matlab/Simulink simulation model. This work is part of the SETTA [4] project (Scheidler *et al.*, 2000), which addresses the design of fault-tolerant and safe distributed computer control systems in the area of future automotive, aerospace and railway systems. The WCET analysis is integrated into the software engineering process of the SETTA project. This software engineering process is designed to automate the simulation and rapid prototyping of dependable real-time computer systems. The focus of this project lies on time-triggered computer systems (Scheidler *et al.*, 1997).

The communication nodes of systems can be modelled with Matlab/Simulink and verified with the WCET analysis framework, presented in this work. This allows the generation of timing information automatically by generating high-level language code from abstract application models. It further allows taking into account optimizations performed by the compiler. This framework eases the design and analysis of distributed dependable computer systems.

The paper is structured as follows: Section 2 gives an overview of the main components of the WCET analysis tool chain. The concept of the WCET analysis method and tool is given in Section 3. Section 4 introduces the components that are used for the simulation model and the method used to integrate the generation of program path annotations into the target source code. Section 5 describes the integration and interfacing methods of the model management and the WCET analysis. Finally, Section 6 presents conclusions and plans for future work.

2. WCET FRAMEWORK

The WCET analysis framework chain can be decomposed into two main modules as shown in Figure 1.

---

[4] SETTA ... Systems Engineering for Time Triggered Architectures

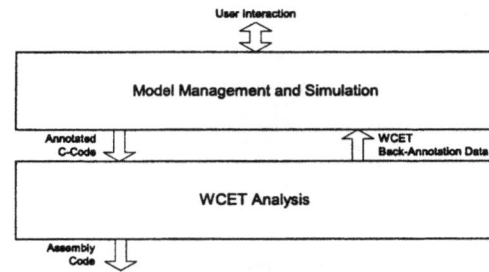

Fig. 1. Main Modules of the WCET Analysis Tool Chain

The first module, denoted "Model Management and Simulation", is responsible for obtaining all data relevant for the simulation model from the user and maintains this data for further processing. In addition, this module provides model simulation to allow, for example, designing controllers for open or closed-loop control. The module further provides code generation, to produce executable algorithms of designed and analyzed controllers.

The second module, denoted "WCET Analysis", calculates the WCET of an executable algorithm that is provided in the form of annotated C code by the "Model Management and Simulation" module. The "WCET Analysis" module is self-contained and operates autonomously. For reasons of transparency towards the user, however, the user interaction with the "WCET Analysis" module is performed via the "Model Management and Simulation" module. The results of the WCET analysis are propagated back to Model Management in the form of WCET back-annotation data. The WCET analysis process is directly coupled with code generation.

3. WCET ANALYSIS TOOL

The task of the "WCET Analysis" module is to derive the WCET from the annotated C code and to provide a correlating assembly output that can be executed on the target hardware. In order to derive safe upper bounds, the WCET analysis method is based on "Static Runtime Analysis". The theoretical concepts of this approach are described in (Puschner and Koza, 1989; Puschner and Schedl, 1997).

The WCET analysis is performed by gathering timing information at a high representation level of the application (simulation model) and performing WCET analysis at assembly-language level. This allows the generation of the timing information automatically by the target-language compiler of the model management module (see Section 4 for further details) when emitting the program code for the model. The annotated

C code acting as a high-level programming interface is derived from ANSI C. Besides the standard C statements it has several extensions to express information about the control flow inside the source file. This language is called WCETC (see Section 5.1).

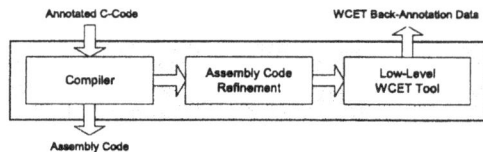

Fig. 2. WCET Analysis Tools

The WCET analysis tools process the annotated C code and produce a result file containing the calculated WCET values and an assembly file that can be processed by an assembler to generate a corresponding executable. As given in Figure 2, the schematic structure of the WCET analysis tools consists of three main components:

Compiler: The compiler is derived from the GNU GCC [5] using the ANSI C language parsing front end, and modified to translate annotated C code.

Assembly code refinement: In this stage, the refinement from assembly code containing abstract instructions to the specific code is performed (e.g., to a specific address distance for a generic branch-instruction).

Low-level WCET tool: This tool computes the WCET bounds. This information is propagated back to the "Model Management and Simulation" module.

In the following the functionality provided by the compiler and the low-level WCET tool are described in more detail.

3.1 Compiler

The chosen approach of WCET analysis uses timing annotations inside the high-level programming language and performs WCET calculation at assembly level. The compiler transforms the information about program structure and WCET timing information. This transformation is done correctly, even in the presence of compiler optimizations. This approach is described further in (Kirner, 2000).

The grammar of the compiler has been extended to accept the WCET annotations for the programming language and to transform them correctly up to the assembly code generation phase.

3.2 Low-Level WCET Tool

The low-level WCET tool is the core component for calculating the WCET of a program. This tool processes the refined assembly code input files that are produced by the compiler.

To reduce the time-consuming method required for searching the program flow path through the structure tree of a program, integer linear programming (ILP) is used. The WCET calculation is performed directly on the solution of the ILP problem.

4. MODEL MANAGEMENT AND SIMULATION

The "Model Management and Simulation" module is based on the off-the-shelf available Matlab/Simulink environment. The main components of this module are given in Figure 3. This section gives a short introduction of these components and a description about how to apply the WCET analysis concept to the simulation model.

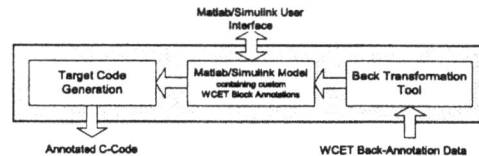

Fig. 3. Model Management and Simulation

4.1 Required Adaptions

Matlab/Simulink is the core of the model simulation environment. The code generation for the simulation models is done by using the Target Language Compiler (TLC) from the Real-Time Workshop (RTW) which is an extension to Matlab/Simulink for rapid prototyping, real-time simulation or stand-alone simulation.

The Matlab/Simulink environment has to be extended to provide custom WCET annotations for the blocks used in an application. In addition, the blocks also have to be able to store the results of the WCET analysis.

For the target code generation stage, the TLC was configured to generate annotated C code (see Section 5.1 for further details).

The required modifications and extensions of the Matlab/Simulink environment can be summarized as follows:

- Provision of custom WCET block annotations within the Matlab/Simulink environment.

---

[5] GCC ... GNU Compiler Collection

- Design of a verification tool to verify the custom WCET block annotations prior to the code generation stage.
- Modification of the configuration files for the TLC to account for the block annotations and to produce correctly annotated C code.
- Design of an appropriate back-transformation tool to reintegrate the WCET results into the Matlab/Simulink model.

4.1.1. *Assumptions and Limitations*

Due to the requirements for enabling WCET analysis of the generated code, several limitations have to be applied to the generic Matlab/Simulink simulation model:

4.1.1.1. *Single Rate Limitation*

Simulink simulation models can contain from one up to several sample rates. In case of a multi-rated Simulink model (with two or more sample times) the WCET analysis of the synthesized code has to deal with several aspects that are difficult to handle, e.g., floating point context switches, operating system delays, preemptive execution, etc.

Therefore, the model management is restricted to single-rated models, where these effects can be avoided.

4.1.1.2. *Bare-Board Target Assumption*

The bare-board target assumption requires the execution of the model in the context of an interrupt service route, with no operating system environment.

This approach was chosen for the generated code to avoid interferences of a specific real-time operating system with code execution.

4.1.1.3. *Embedded-C Code Format Assumption*

The code generation is derived from the embedded-C code format of the embedded real-time target. This code format is optimized for speed, memory usage, and simplicity. Since this format is intended for use in embedded systems it was chosen as the base for WCET analysis.

4.1.1.4. *Integral Data Type Blockset Limitation*

Since most processors do not support floating point operations, special library function calls would be required to handle them. To avoid any loss of accuracy for the calculated WCET bound, only blocks that do not lead to floating point code are allowed in the simulation model.

4.1.1.5. *Limited block subset*

Due to the great amount of available TLC configuration files, only a subset of the blocks has been adapted to produce WCETC compliant code. The following categorization of the supported blocks can be made according to their impact on the produced code:

No impact: Subsystems and their signal handling blocks (in and out blocks) are used to provide a hierarchical information abstraction of the model to the user. They have no impact on the produced code. Therefore, these blocks are compliant to the WCET analysis method without modifications or restrictions.

Non Executing Code: Blocks that produce configuration code such as constant values but do not produce runtime executing code, can also have an impact on the WCET of the resulting code. To conform to the WCET analysis method, these blocks do not require modifications, but their usage is restricted to integral data types.

Executing Code: Blocks that produce code that performs calculations during run-time of the program. Since such blocks can lead to code that contains conditional statements (e.g., loop constructs), they have to be modified to generate also annotations about the runtime behavior. For some types of blocks, additional user input is necessary to generate these annotations.

From the *executing code*, only sum, gain, product, algebraic constraint, logical operator, manual switch and the RTW's C custom code model output function block are currently supported.

4.2 *Integrating Information for WCET Analysis*

For the integration of WCET analysis into the simulation model, it was necessary to adjust the TLC to generate C code, which is annotated with execution path information. The target language used to enable embedding of annotations is WCETC. The generated target language by the TLC depends on the TLC file hierarchy which describes the produced code fragments for each simulation block and their connectivity. The required extensions to express path information for WCET analysis have been placed directly into the TLC files.

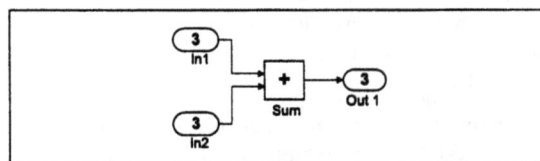

Fig. 4. Simulink model of a vector operation

The required modifications depend on the generated code structure. To visualize this process, an example of a generated code for a simple block is given. This block calculates the sum for each index of two input vectors. The input vectors for this model are restricted to one-dimensional vectors of length 3. The block representation of this model is given in Figure 4.

```
/* Gain Block: <Root>/Sum */
{
  int_T i1;
  int8_T *u0 = &rtB.In1[0];
  int8_T *v0 = &rtB.In2[0];
  int8_T *y0 = &rtB.Out1[0];

  for(i1 = 0; i1 < 3; i1++)
  maximum (3) iterations
  {
    y0[i1] = u0[i1] + v0[i1];
  }
}
```

Fig. 5. Code produced for Simulink model

The input blocks In1 and In2 represent import blocks that receive their values from other blocks. The Sum block calculates the sum of both inputs for each index of the vector. Therefore, the output block is also a one-dimensional vector of length 3 with the same atomic data type as inputs.

At this point it is also important to remark that Simulink has a strict data type checking. Type conversions have to be done by using special conversion blocks. This behavior meets the requirements of not mixing data types of different precisions in the same expression as proposed in (MISRA, 1998) for real-time software in the automotive domain.

The code generated for the example is given in Figure 5. The additional annotations for WCET analysis are shown in bold style. Since the array dimensions and lengths are known statically, the array processing in this example is just transformed into a loop with the vector length as upper loop bound.

This example shows how target code is generated from the simulation model and how path annotations are inserted into the code.

5. INTERFACING THE COMPONENTS

As said above, the WCET framework is composed of two main modules, called "WCET Analysis" and "Model Management and Simulation". This section describes the interfaces between these two modules.

- Simulation Model ⟹ WCET Tool:
  This interface is given by the syntax of the programming language WCETC .

- Simulation Model ⟸ WCET Tool:
  This interface is defined by back-annotation data, which describe the calculated WCET for the blocks of the simulation model.

The interfaces are described more precisely in the following:

5.1 Simulation Model ⟹ WCET Tool

The interface between the Simulink environment and the WCET analysis is defined by WCETC, which can be summarized as a superset from a subset of ANSI C (Kirner, 2000). The following constructs (Puschner and Koza, 1989) have been added to the syntax of ANSI C to enable static WCET analysis:

Loop Bound: Loop bounds describe the maximum iteration count of a loop construct.
Scope: A scope is a local environment for expressing possible program execution paths.
Marker: Markers are used to label a certain position of a program.
Restriction: Restrictions in the form of equations and inequalities are placed at the end of a scope to describe the program control flow of the code inside the scope. These inequalities are used to build the restrictions for a linear programming problem (Puschner and Schedl, 1997), on which this WCET analysis method is based.

An example for a code fraction of WCETC using a loop bound was already given in Figure 5. Beside the integrated annotations for describing (in)feasible program execution paths there are additional annotations to identify the corresponding Simulink blocks:

Block Info: These annotations describe the assignment of code fractions to the corresponding Simulink blocks. This enables the identification of needed execution times at finer granularity of the simulation model.
Build Info: This information is used for code revision tracking. It is generated by the simulation tool and passed through the whole analysis environment. It is not required for the WCET analysis itself.

5.2 Simulation Model ⟸ WCET Tool

The "WCET Analysis" module calculates the WCET from the annotated source code and provides the results at a certain granularity. The back transformation tool processes the WCET back-annotation data. The main task of this tool is to map the WCET information to the associated blocks and to generate WCET information of subsystems that consist of multiple blocks using the WCET information that is available for the respective blocks contained within such a subsystem.

```
INTERFACE_FILE: BUILD_INFO BLOCK_RESULTS
BUILD_INFO: "build(" _string_ ")"
BLOCK_RESULTS: BLOCK_RESULT BLOCK_RESULTS | $
BLOCK_RESULT: C_FILE "," C_FUNCT "," BLOCK_NAME
                "," CYCLES
```

Fig. 6. Interfacefile Simulation $\Longleftarrow$ WCET

The EBNF-grammar[6] for such an Interface is given in Figure 6, where C_FILE, C_FUNCT and BLOCK_NAME are string constants for the corresponding data fields. The calculated WCET for each block is given in CYCLES.

6. SUMMARY AND CONCLUSION

The novel concept described in this work lies in the combination of WCET analysis with model based system specifications. The WCET analysis method is based on static program analysis to enable the calculation of safe upper WCET bounds. The simulation environment supports the design of software and automatically performs a simulation and code generation for it. The generated code is automatically processed by the WCET analysis tool which produces WCET information for the Simulink blocks and also the final executable program.

This framework is well suited for the design of communcation protocols of dependable computer systems, because their behaviour can be simulated directly inside the simulation environment. To guarantee a safe runtime behavior of such a system, the WCET of each program can be calculated directly.

For the future it is planned to demonstrate the possibilities of this framework by several concrete sample applications.

7. REFERENCES

Bernat, G., A. Burns and A. Wellings (2000). Portable Worst-Case Execution Time Analysis using Java Byte Code. In: *In Proceedings of the 6th International EUROMICRO conference on Real-Time Systems*. Stockholm.

Erpenbach, E. and P. Altenbernd (1999). Worst-Case Execution Times and Schedulability Analysis of Statecharts Models. In: *Proceedings of the 11th Euromicro Conference on Real Time Systems*. York.

Harel, D. and A. Naamad (1996). The STATEM-ATE Semantics of Statecharts. *ACM Transactions on Software Engineering and Methodology (TOSEM)*.

Harmon, M., T. Baker and D. Whalley (1994). A Retargetable Technique for Predicting Execution Time of Code Segments. *Real-Time Systems Journal* **7**(2), 159–182.

Kirner, R. (2000). Integration of Static Runtime Analysis and Program Compilation. Master's thesis. Technische Universität Wien. Vienna, Austria.

Klingerman, E. and A. Stoyenko (1986). Real-Time Euclid: A Language for Reliable Real-Time Systems. *IEEE Transactions on Software Engineering* **12**(9), 941–989.

Li, Y.-T. S., S. Malik and A. Wolfe (1995). Efficient Microarchitecture Modeling and Path Analysis for Real-Time Software. In: *Proceedings of the IEEE Real-Time Systems Symposium*. pp. 298–307.

MISRA, The Motor Industry Software Reliability Association (1998). *Guidelines For The Use Of The C Language In Vehicle Based Software*. ISBN 0-9524156-9-0. MIRA.

Park, C. Y. (1993). Predicting Program Execution Times by Analyzing Static and Dynamic Program Paths. *Real-Time Systems* **5**(1), 31–62.

Puschner, P. and A. V. Schedl (1997). Computing Maximum Task Execution Times – A Graph-Based Approach. *The Journal of Real-Time Systems* **13**, 67–91.

Puschner, P. and C. Koza (1989). Calculating the Maximum Execution Time of Real-Time Programs. *The Journal of Real-Time Systems* **1**, 159–176.

Scheidler, C., G. Heiner, R. Sasse, E. Fuchs, H. Kopetz and C. Temple (1997). Time-Triggered Architecture - (TTA). *In Advances in Information Technologies: The Business Challenge, IOS Press* pp. 758–765.

Scheidler, C., P. Puschner, S. Boutin, E. Fuchs, G. Gruensteidl, U. Virnich, Y. Papadopoulos, M. Pisecky and J. Rennhack (2000). Systems Engineering of Time-Triggered Architectures - The SETTA Approach. In: *In Proceedings of the 16th IFAC Workshop on Distributed Computer Control Systems*.

Vrchoticky, A. (1994). Compilation Support for Fine-Grained Execution Time Analysis. In: *In Proceedings of the ACM SIGPLAN Workshop on Language, Compiler and Tool Support for Real-Time Systems*. Orlando FL.

Zhang, N., A. Burns and M. Nicholson (1993). Pipelined Processors and Worst Case Execution Times. *Journal of Real-Time Systems* **5**(4), 319–343.

---

[6] EBNF ... Extended Backus-Naur Form

Copyright © IFAC Distributed Computer Control Systems,
Sydney, Australia, 2000

REAL-TIME BEHAVIOUR VERIFICATION, ANIMATION AND MONITORING STARTING FROM DCCS SPECIFICATION

L.Motus[*] and T.Naks[*#]

[*]*Department of Computer Control, Tallinn Technical University and [#]IB Krates*
leo.motus@dcc.ttu.ee, tonu.naks@ttu.ee

Abstract: This paper describes efforts made to facilitate the use of UML based tools (e.g. Booch et al, 1999) in time-sensitive and dependable distributed applications. Those efforts are based on theories developed and applied for analysing timing properties of time-constraint concurrent software. They also include development, testing and (partial) integration of a software engineering tool (dedicated to formal timing analysis of software) to the suite of UML based tools. The resulting complex is under test (in Systems and Software Engineering Laboratory at Tallinn Technical University), separate tools have been tested in two European Union research projects (CP-94 no. 1577 LIMITS, and ESPRIT 22154 BRIDGE). The usefulness of presented ideas and results are based on consequences that result from introducing a more complex time model into real-time software engineering.

The approach described in this paper suggests a practical tool in support of the idea (Xu and Parnas, 2000), that the role of (priority) scheduling is overemphasised in the practice of software development. The most effective stage for starting timing analysis is the specification. The most salient timing errors can, if suitable formalism (e.g. Motus and Rodd, 1994) is used, be discovered already before the design stage starts. This approach assumes, before the specification can be completed and analysed for timing correctness, a thorough systems' analysis and some understanding of the control theory that is to be used. *Copyright © 2000 IFAC*

Keywords: Real-time behaviour; DCCS lifecycle; formal verification of timing properties; automatic animation of a specification; a dedicated CASE tool integrated with UML based toolkit.

1. INTRODUCTION

Distributed computer control systems have become inseparable part of our everyday life although many aspects of their specification, design, and implementation are, to a large extent, based on empirical experience and subjective decisions. As an example, specific timing and dependability requirements to distributed control system's software and also rather strict, deterministic performance requirements to the computing and communication media are still difficult to analyse with necessary rigour – especially at the early stages of software development.

Only too often the timing issues are seriously dealt with at the end of physical design stage and at the implementation stage. In the likely case that some timing (or scheduling) inconsistencies will be detected in system's real-time behaviour, the required changes are pretty expensive to introduce. In not too rare occasions the inconsistencies will be discovered only when the system is already operational. For instance, disturbances caused by saw-tooth behaviour of a message delay in asynchronous communication (see Motus and Rodd, 1994, ch.7) are very seldom detected by testing. Another example of software timing problems is discussed in (Albertos and Crespo, 1999) where the inconsistency is caused by excessive delay in output of control variable value, or by irregular (non-uniform) sampling of measurements. In both cases "excessive delay" and "irregular sampling" means that due to software design and/or implementation the actual delays and irregularities do not meet expectations of the used control theory.

The most recent arguments in favour of the earlier start of studying real-time behaviour, including quantitative timing and scheduling are given in (Xu and Parnas, 2000). This paper provides additional practical weight to those arguments by discussing experience of using a CASE tool LIMITS (based on theory developed in (Motus and Rodd, 1994)) that is linked to UML based CASE tools. This CASE tool supports many pre-run-time scheduling issues, e.g. it analyses quantitative timing properties, co-ordinates synchronisation precision of activities, enables to estimate and/or calculate traffic intensity in communication network, provides analytical mini-max estimates to expected performance characteristics. The first estimates can be received from the specification. As the system develops, the estimates can be updated so as to match with the actual properties of the design and/or implementation.

The UML (Unified Modelling Language) is chosen because of its maturity, its wide acceptance by users from different user communities and good support by a variety of CASE tools. From the DCCS software point, in spite of all its virtues, the UML is not sufficiently effective in describing and analysing quantitative timing properties of interactions, and sequencing. Requirements on validity time of data and events, and precision of synchronisation of various activities are also difficult to handle in UML.

To compensate these deficiencies a new elaborated process view is introduced (in the UML terminology). The new process view is presented by a formalism (the Q-model) suggested in (Motus and Rodd, 1994) and supported by a CASE tool LIMITS. The new process view is based on information, contained in UML static (class and object diagrams) and dynamic (behavioural) diagrams plus additional required information related to the requirements from the application and from the involved control theory. A more detailed example of how to form the new process view (Q-model) is given in (Motus and Naks 1998).

Problems to be solved at different lifecycle stages of a DCCS. Normally a DCCS is being developed (evolving) over a remarkable period of time – it is designed, implemented, and modified step by step. It definitely helps if a master plan for the final product exists. Quite often, however, it is not possible to

forecast time of the following development step. Therefore integration of various parts of a DCCS, developed at different time and maybe based on different technology, is not a trivial task.

The suggested verification, animation and monitoring tool/method is also useful for detecting potential inconsistencies when integrating separately developed subsystems, but this needs a separate paper. In the following three basic DCCS development stages are considered and tasks solvable by the suggested tools are briefly characterised. The main emphasis is on early processing of quantitative timing information.

2. SPECIFICATION AND ARCHITECTURAL DESIGN

DCCS software specification and design needs somewhat special handling (as compared to that of information processing system) mainly because of the complexity of interrelations between requirements to computer system from the controlled object, and numerous intrinsic non-functional requirements of the computer system. In some cases this leads to serious theoretical problems – for instance, forced concurrency of software (Motus and Rodd (1994), or variety of simultaneously used time concepts (time models) required to adequately describe systems real-time behaviour (Motus 1993).

This stage comprises, in addition to UML prescribed traditional activities that are focused mainly on functional requirements, also specification of components responsible for satisfying non-functional requirements. Typical non-functional requirements for a DCCS are, for instance:
- quantitative timing constraints, ordering and synchronisation requirements imposed by controlled object and the human operator
- quantitative timing constraints, ordering and synchronisation requirements imposed by (or deducible from) the soft- and hardware components of the system.

In practice, such information is often added at the component implementation stage, or even immediately before the integration and testing stage. Conventional argument is that the designer does not know sufficiently precise estimates for the respective system characteristics. This widely used argument is true only partially and may result in excessive amount of errors that are to be detected at the integration, schedule compilation and final testing stages. In addition to the high price of rewriting substantial amount of code to eliminate the detected errors, such practice cannot – in principle – detect subtle timing errors that possibly stem from specification and design stages.

In the majority of cases, when speaking of timing constraints, the designers mean performance characteristics – such as, execution time of a task (processor time required by a task), task termination deadline, task activation deadline, response time for a sequence of tasks, and others.

Very seldom, in practice, are constraints set on the validity time, of a value assigned to a variable, or that of an event. Even more seldom are the designers interested in timing properties of inter-process communication that are closely related to validity time of exchanged messages and to the synchronisation precision of invoked events (or activities). This practice is still continuing, although the analysis of many accidents points to the out-aged (or inconsistent) variable values presented to human-operators as the major cause of misjudgements that may have lead to accidents, see, for example (Bransby, (1998)).

Computer science relies heavily on incompletely constraint synchronisation. For instance, the termination event of computing process A activates two or more other processes (e.g. B, C, and D). The end user normally states that simultaneously with the termination of process A, processes B, C, and D are activated. Programmers say that processes B, C, and D are executed in parallel, and their execution is invoked by the termination of process A. In many cases maximum delay between the termination of process A and activation of B, C, and D is specified.

Very seldom, however, is the simultaneity interval for B, C, and D activation instants specified. In some cases the simultaneity of B, C, and D activation instants does not really matter. In the cases where it does matter, the common answer is – this can properly be fixed at the implementation stage, when the other scheduling problems are solved. This suggestion is based on an unduly optimistic assumption that any synchronisation precision problem can be resolved at implementation and integration stages of the project. In many applications the synchronisation precision is vital for the project – e.g. car's or plane's anti-sliding brakes. It would be safer for the project to consider the required synchronisation precision from the specification stage, and apply specification, design and implementation methods that together with properly selected hardware configuration can guarantee the required precision.

Schedulability can often be achieved only by "slightly" modifying the earlier specified timing requirements. However, many timing constraints, especially those imposed by a physical process that takes place in the controlled object, cannot be modified. The software designer can only modify constraints that are strictly intrinsic to computer system. At the same time the controlled object determines, to a large extent, the behaviour of

software in computer control systems. The designer should make a very clear distinction between time constraints that are "frozen" by the controlled object and constraints whose values can be modified, if necessary. Also, many constraints are fixed heuristically, or based on expert knowledge, and consequently they are not necessarily consistent with each other.

Another difficulty for schedulability could be the dynamically changing number of copies of a computing process – essentially caused by forced concurrency and the required paradigm for formal description of a computing process in this mode (see, Motus (1995)).

Conventional software engineering leaves such decisions to the designer and does not enable sufficient support to consistency analysis of various time constraints and requirements on synchronisation precision.

2.2 The role of time models in handling software real-time behaviour

The software designers' wide spread self-limitation to studying only performance-bound characteristics has two causes:
    - for not very demanding applications the performance-bound characteristics give adequate description of the systems behaviour
    - commercially available CASE tools cannot, in principle, support the analysis of more sophisticated time characteristics because of the trivial time models supported by them.

Motus (1993) demonstrates that three different philosophical concepts (models) are required to describe and analyse all the software timing aspects. These three concepts are – non-reversible (or thermodynamical) time as in biology, fully reversible time (as in theoretical physics), and relative time (as in psychology) where the origin of time is always in present (see, Denbigh (1986)). How those different aspects of time are used in formal analysis of timing correctness is described, for example in Motus and Rodd (1994), Motus (1995).

A time model, typically used in commercially available CASE tools, is completely reversible global time – the behaviour of all the computation processes can be projected to a single time axis and be described in the same time. This sounds pretty attractive but does not reflect the reality too well – in one DCCS the time scale of processes may reach from milliseconds to months. It is more appropriate to have separate time attached to each process – analogously with the quantum physics.

For error recovery time, in a typical time model, can be reversed whenever necessary. This does not give a true picture of the real world, where time is not fully reversible. Time can be measured in natural numbers or in real numbers. In some models, especially in less demanding tools and/or methods, time may have no metric properties, i.e. time is topological and can only describe relative order of events (Motus, (1993)). Fully reversible topological time has been the most popular time model in computer science.

Time model that uses real numbers provide a dense time – between any two neighbouring time instants one can insert infinitely many time instants. This is a huge advantage from the philosophical point of view. However, from the practical point of view such time cannot adequately be implemented in a computer. Processes modelled in continuous time can, not so seldom, be considered only as approximations of real world processes. This situation is analogous to difficulties of substituting continuous time based mathematical theory with discrete time based theory – this cannot be always done without loss of some properties.

One may even have difficulties with describing and/or detecting causality in a continuous time theory (see, for example Caspi and Halbwachs (1986))

2.3 A formalism for analysing time behaviour of DCCS software

The arguments provided in the previous sections explain the necessity of developing a more advanced, and not too sophisticated formalism that would enable some formal analysis of the system's behaviour, based on the incomplete specification of a DCCS. Preferably the analysis should become better step-by-step, as the specification becomes more complete and the estimates of parameter and constraint values improve.

Three basic features of conventional software models that had to be changed in order to obtain a suitable formalism (Motus and Rodd (1994)) are:

- A rather popular but idealistic paradigm (real-time software is a non-terminating program) had to be changed to a more pragmatic one -- a real-time software is collection of repeatedly activated, terminating, and interacting with each other, computing processes; this modification enables, in principle, formal verification of real-time software;
- Each single computing process may be activated repeatedly (possibly countable number of times), producing each time new values for its output variables; since activation instants of interacting processes need not be synchronised, it becomes important to differentiate between the results obtained at different executions -- i.e. a process is to be described as a specific mapping

$p: T(p) \times \text{dom } p \rightarrow \text{val } p$, where $T(p)$ is the set of activation instants;

- Interaction between any preliminarily fixed executions of the two processes is allowed, provided that the partners can be uniquely identified and the satisfaction of imposed time constraints can be checked. This introduces a time-selective inter-process communication that is not always comparable to conventional queue and stack based communication; time selective communication is often necessary in control applications (e.g. operator information system at alarm processing).

There exist at least two formalisms that satisfy the above stated new features:

- a process algebra approach (Caspi and Halbwachs, (1986)), based on continuous time; it leads to huge computational difficulties, mostly due to the use of continuous time, there is no information about the CASE tools supporting this approach;
- seemingly empirical Q-model (Motus and Rodd (1994)), based on discrete time that can, however, be presented as a weak second-order predicate calculus; the Q-model has not been too widely accepted, although formalism can effectively be hidden by a CASE tool LIMITS

2.4 Analysing capabilities of the Q-model

Each process is assigned with a number of timing parameters – such as, activation instants, execution time, and others. Timing requirements or constraints are also assigned to pairs of interacting processes – e.g. acceptable message delay, validity time of the message, instant of interaction. Synchronisation precision should be fixed whenever necessary. At the specification stage all the values of those timing parameters and constraints are estimates, or requirements derived form properties of the controlled object. As the system's design evolves, the initial estimates are substituted by more advanced estimates and the analysis is repeated.

Based on the assigned constraints and parameter values, many timing properties of the specification and/or design can be checked analytically – see, for details Motus (1995). All the assigned estimated values may be given as min-max intervals. This feature of the Q-model is different from the majority of run-time schedules that assume fixed value of time parameters. Another difference from run-time scheduling is that strictly periodic processes can be handled together with sporadic activation periods.

Many interesting by-products have emerged during analytical behaviour study of real-time software. For instance, messages exchanged between two asynchronously executed processes have a non-transport delay that changes from execution to execution as a saw-tooth graph. It is impossible in such a communication to detect and eliminate the violation of maximum acceptable delay via testing. Another interesting feature is the possibility to calculate maximum number of simultaneously executed copies of the same process, required to satisfy the forced concurrency requirements from the controlled object. (Motus and Rodd (1994)).

Only a small part of Q-model provided capabilities has been studied so far – not many applications require such depth of pre-implementation analysis in a present day world.

3. CO-OPERATION OF LIMITS AND UML BASED TOOLS.

Any formalism used to support software development should be effectively hidden from the end user before its wide acceptance can be expected. A large part of Q-model properties and analysing capabilities is implemented in a CASE tool LIMITS. The tool automates many routine procedures, and relieves the end user from the details of the underlying formalism. Nevertheless, timing correctness forms only a small part of system designer's worries. Therefore it is too naive to hope that such a narrowly focused CASE tool will be widely accepted – unless the tool co-operates with some universal software design tool. UML based tools have been selected as the most elaborated and widespread.

The co-operation of the two tools is organised on mutual benefit. UML based tool is used for compiling a general description about the future system – elaborated methodology and semiformal nature of UML has facilitated its wide acceptance. The information collected into UML diagrams is used, and missing time-specific information is added, in the process of building a Q-model diagram in LIMITS. The Q-model is automatically analysed for inconsistencies and correct time-wise behaviour. When the model is formally correct, the designer may wish to do informal analysis – test the specification on a separately specified scenario. The animation results can be visualised automatically and used for modification of the model. Then the formal analysis has to be repeated. Finally, the verified Q-model may be used manually to update respective UML diagrams – direct addressing link is maintained between UML class diagram and the Q-model. The other diagrams are to be modified based on indirect information transfer manually.

3.1 Basics of LIMITS

LIMITS tool runs on Win/NT platform, requires minimum 64MB of memory, and 10MB hard disk plus storage space for project data. In principle it is a stand-alone CASE tool, although its inference engine has been used separately for OEM applications (Naks and Motus (1999)), and the tool itself has been linked to OMT/StP and Artisan RTS. LIMITS tool has a conventional GUI, a Q-model diagram builder, repository, analyser, and animator together with the animation visualiser.

The user describes the system in the Q-model notation, specifies the estimated timing parameter values, constraints, and requirements. The syntactic correctness of the input information is checked on-line. Timing correctness is analysed in three steps:

- correctness of single processes
- correctness of interacting pairs of processes
- correctness of group behaviour

After the specification's behaviour has been proven formally correct, the user may initiate informal analysis (animation of the specification under user described scenarios). The prototype is automatically generated, and the user may interactively supervise the animation. After the animation is complete, the resulting time diagrams can be separately stored, visualised and analysed manually or by a user written program. The animation diagrams display the concurrent execution of selected processes, with specially marked interaction sessions, and violations of specified constraints.

Not all the theoretically feasible capabilities of the Q-model have been implemented in LIMITS. At the moment LIMITS is regularly used by students for practical exercises in real-time software engineering at Tallinn Technical University.

3.2 Why is co-operation of LIMITS with a UML based tool important?

UML based tools support systems description from many viewpoints – the resulting description is presented with a variety of models (one can choose from nine basic diagrams, depending on the complexity and nature of the future system). There is no doubt that UML is, in practice, the most thoroughly elaborated modelling methodology (see, for instance, Booch, Rumbaugh, and Jacobson (1999)).

However, UML is developed as a universal modelling language that has to cover the needs of a wide variety of applications. Distributed computer control systems, especially those with hard time constraints, do not form, yet, the overwhelming majority of computer applications. Hence, handling timing correctness of the behaviour is, at the moment, not the strongest feature of UML.

On the other hand LIMITS is pretty well focused on handling timing correctness issues at different stages of systems' development. It seems a natural idea to use Q-model (as implemented in LIMITS) as one of the many complementary models of the UML. In UML terminology the process view is the closest to what LIMITS actually supports. The process view encompasses threads and processes that form the system's concurrency and synchronisation mechanisms. Performance, scalability and throughput are addressed by a process view. The other views listed in (Booch, Rumbaugh, Jacobson (1999)) are – use case, design, implementation and deployment views. Two UML models (diagrams), namely class diagram and interaction diagrams are recommended for creation of the process view. The interaction diagram is usually split into two – a sequence diagram and a collaboration diagram.

For the real-time behaviour verification the difficulty with many UML diagrams is that none of them is formal enough to enable true verification of timing properties. Besides, the useful timing verification information is distributed among several diagrams, with still part of the information missing. One can find useful bits of information, for example, from class diagrams, use cases, integration diagrams (sequence and collaboration), statechart diagrams, activity diagrams. By definition, different UML diagrams are rather independent from each other. This feature is useful for manual detecting of missed details or features, by comparing different views on the reality and empirical understanding of the reality by a group of designers. However, that feature also makes automatic consistency checks of overlapping information that is presented in different diagrams practically impossible and causes huge difficulties when trying automatically to generate a prototype based on a UML described specification or design.

3.3 How LIMITS and UML based tools co-operate?

There are many choices for co-operation of CASE tools – from complete integration to real loose collaboration. The authors have pragmatically selected the latter option – LIMITS and UML based Artisan Real-time Studio collaborate via one-way link that enables to relate a class diagram and its classes to the components of Q-model diagram in LIMITS. All the other useful information (i.e. the other UML diagrams) should be accessed manually. The feedback from the verified and time-wise correct Q-model to the UML diagrams – if the modification of the UML diagrams is considered necessary – should also be done manually. Although such collaboration is rather primitive it provides useful support for the users of both tools and the navigation between the two tools is rather straightforward.

The LIMITS users can get access to well organised and in many aspects well checked description of the future system. It is true that the description does not provide all the data required for complete analysis of real-time behavioural correctness of the system. The missing data is added in the process of Q-model diagram compilation. At the same time, correspondence between the classes, operations, and associations from the UML class diagram and processes, channels, and clusters from the Q-model diagram is established. For the missing data and parameter values, the designer has to browse the other UML diagrams, perform additional tests on the controlled object, or elicit expert knowledge from specialists.

When the Q-model description is completed, the formal analysis has to be passed first and only then animation on an automatically generated prototype can be performed. In not so rare cases the UML class model and some other UML diagrams are to be changed to accommodate the modifications made to pass the formal analysis, or to satisfy the users expectations during the animation. As an extra gain, a set of quantitative time constraints (mostly on the software, but also on the hardware configuration and network protocols) are obtained that should be satisfied during coding, schedule building, and system integration.

3.4 Some comments on the UML real-time extension, as planned today

The authors wish to compliment the OMG for sending out the RFP and especially for excellent mandatory and optional requirements in the RFP. The same applies to the authors of a response (OMG document ad/2000-08-04) for the thorough and excellent work done.

Alas, nobody has seen absolute perfection. The authors of this paper wish to emphasise two aspects of the proposal that are too heavily based on conventional (mainstream) understandings of today and might harm further evolution of the UML real-time extension. The primary concern is related to the limited selection of time models. The second concern is less crucial and is related to insufficient attention to pre-tun-time scheduling.

Basic time related problem is the assumption that time can only progress monotonically – this may be valid outside of the computers, but is definitely not valid inside. If one does not allow a multitude of times (including reversible time and relative time with moving origin) the analysis is limited to performance-bound properties only. All the truly salient time inconsistencies and constraints occurring during time-selective inter-process communication will be neglected, in principle. See also section 2.2 of this paper.

One should not forget what happened with time in physics – the transition from one global monotonic (and reversible) time in the 1920-s to quantum theories where each particle has its own time. Analogously, only the multitude of times enables the modeller to analyse the behaviour of each class (object, operation) separately from other model elements, and in interaction with the other model elements.

Scheduling has proven to be important and useful in practice. However, quite often in real-time applications the execution of many tasks is controlled by the events and physical processes that occur outside of the computer system, and are out of the modeller's influence. Conventional scheduling of software tasks in a computer is similar to open-loop control in automatic control theory. It has been demonstrated that feedback control is more effective than open-loop control in incompletely known situations. The same statement translated into computer science language would claim that pre-run-time scheduling might have a much larger role in real-time software development than generally believed today. Pre-run-time scheduling operates at specification and design stages when a mixture of software and parts of the control object are handled together.

4. PHYSICAL DESIGN AND HARDWARE TOPOLOGY

The UML books call activities of this life cycle stage architectural modelling. When in specification and logical design stages one was interested in logical entities (classes, associations, interactions, statecharts, etc), then in physical design (architectural modelling) stage the main interest is on software components. A component is a physical and replaceable part of a system. The UML provides component and deployment diagrams to describe the physical design of the system. The component diagram shows the organisation and dependencies among a set of components. The component diagram is then mapped to selected hardware configuration -- so one reaches the deployment diagrams.

The component and deployment diagrams provide conceptual and structural clarity about the topological structure of the future system. However, many behavioural aspects of the future system still remain to be guessed. For instance, the network traffic intensity pattern under varying load conditions. It is important to guarantee maximum acceptable message delivery time under the peak load conditions – typically resulting from the alarm shower. One of the factors that increased the probability of the accident described in (Bransby (1998)), was excessive increase of message delivery delays in a system. This lead to invalid data displayed to operator, and this in its turn eventually caused an inadequate decision made by the operator.

The Q-model view on the software enables, before the software is implemented, to estimate the time-pattern of inter-node traffic, including time distribution of the worst-case traffic loads. This capability is not yet included into the LIMITS tool, but preliminary experiments have been made. The obtained estimates tend to be on the pessimistic side, but not too unrealistic.

Another useful feature of the Q-model at this life cycle stage is that it states explicitly the required precision of synchronisation of events and activities (inside the computer network as well as with the physical processes executing in the environment). Therefore the developer can select a well-founded clock synchronisation precision in the network instead of relying on a gut-feeling or conventional beliefs.

These features are important for appropriate selection of communication media and protocols in a DCCS. The study of expected traffic load, its impact on message delivery times and achievable synchronisation precision quite often forces to change the physical design decisions.

5. RUN-TIME MONITORING OF REAL-TIME BEHAVIOUR

Unforeseen situations in the environment and malfunctioning of soft- and/or hardware may take place during system's operation, even if the design and implementation contains no errors and no inconsistencies. In order to avoid serious mishaps, a run-time behaviour monitoring system is needed in dependable systems. This monitoring system is an extension of an exception handling system and has to cater for correct real-time behaviour. Many monitoring functions can be assigned to run-time schedulers, however, some vital features in dependable real-time systems can not easily be controlled by the conventional schedulers.

Such features are conventionally not related with the scheduling. For instance, inter-process communication restrictions, terminations order violation of simultaneous copies of a process, excessive message delay in communication between asynchronous processes, and others. In such cases one needs a reference value (or interval) that enables to classify the actual value of a parameter as correct or incorrect.

A stand-alone part of a LIMITS tool – the analyser – has been used as a source of reference values and a classifier for run-time monitoring of real-time behaviour in a diagnosing system (Naks and Motus (1999)). The major task of the monitor is to check that the pre-fixed time characteristics of software are actually adhered to during system's operation. The monitor invokes in the case of a violation, whenever feasible, alternative thread of activities.

6. CONCLUSIONS

The complexity of real-time applications increases rapidly, together with the rigidity of dependability constraints. At the time, methodologies and tools supporting the development of systems evolve quietly, instead of a few little revolutions in conventional thinking. This has seriously increased the gap between the complexity and dependability requirements of new systems and the capabilities of an average real-time system developer to match those requirements.

Remarkable progress has been made in the area of object-oriented modelling and design of systems – just imagine the recently emerged UML language and related methodology. On this optimistic background it is especially worrying the self-limitation of real-time system modellers to performance bound properties, and inability (or unwillingness) to cope with more salient timing problems.

The main goal of this paper is to demonstrate the gap between the necessity of real-time systems and between the capabilities of the existing and developed tools. The paper provides some constructive examples that could reduce that gap. For instance, application of sufficiently sophisticated time models, and wider application of the pre-run-time scheduling ideas might provide exit from the "ideological crisis" of software engineering tool developers.

REFERENCES

Albertos, P. and A. Crespo, (1999). Real-time control of non-uniformly sampled systems. *Control Engineering Practice*, vol.7, 445-458

Booch, G., J. Rumbaugh, I. Jacobson, (1999). The Unified Modeling Language User Guide. *Addison Wesley Longman Inc*, 482pp

Bransby, M. (1998). Explosive lessons. *IEE Computing and Control Engineering Journal*, vol.9, no.2, 57-60

Caspi, P. and N. Halbwachs (1986). A functional model for describing and reasoning about time behaviour of computing systems. *Acta Informatica*, vol.22, pp.595-627

Denbigh, K.G. (1981). Three concepts of Time. *Springer Verlag*, 180 pp.

Motus, L. (1993). Time concepts in real-time software. *Control Engineering Practice*, vol.1, no.1, 21-33

Motus, L. and M.G.Rodd (1994). Timing analysis of Real-time Software. Elsevier/Pergamon, 212pp

Motus, L. (1995). Timing problems and their handling at system integration. in Tsafestas S.G. and H.B. Verbruggen (eds) *Artificial Intelligence in Industrial Decision Making, Control and Automation,* chapter 3, 67-88, Kluwer Academic Pulishers

Motus, L. and T.Naks (1998). Formal timing analysis of OMT designs using LIMITS. *International Journal of Computer Systems Science and Engineering*, vol.13, no.3, 161-170

Naks, T. and L.Motus (1999). Handling timing in time-critical reasoning system – a case study. *Proc. IFAC Symposium on Artificial Intelligence in Real-time Control*, Elsevier Science, 1-11

OMG document ad/2000-08-04 version 1.0

Xu, J. and D. L. Parnas (2000). Priority Scheduling versus Pre-run-time Scheduling. *The International Journal of Time-Critical Computing Systems*, vol.18, no.1, 7-23

Copyright © IFAC Distributed Computer Control Systems,
Sydney, Australia, 2000

REFINEMENT AND EFFICIENT VERIFICATION OF SYNCHRONOUS PROGRAMS (EXTENDED ABSTRACT) [1]

S. Ramesh

*Department of Computer Science and Engineering,
Indian Institute of Technology,
Bombay - 400 076, INDIA
e-mail: ramesh@cse.iitb.ernet.in*

Abstract:
Synchronous paradigm exemplified by languages such as Esterel and Statecharts has been found to be useful for programming reactive applications. In this paper, we define a notion of refinement for the synchronous paradigm satisfying the following: A concurrent or hierarchical program refines its syntactic subcomponents and if a program has a property then any program that refines it also has the same property. These suggest a simple compositional verification strategy: in order to verify a program, verify one or more of its syntactic subprograms. *Copyright ©2000 IFAC*

Keywords: Refinement, Synchronous Programs, Temporal Logic, Verification

1. INTRODUCTION

Refinement, well-studied in the context of sequential programs (Back, 1985), (Morgan, 1990) has attracted a lot of attention recently for concurrent systems. In this paper, we study refinement for the synchronous paradigm, introduced in (Berry and Gonthier, 1992), (Halbwachs, 1993), (Maraninchi and Halbwachs, 1996), (Harel, 1987). This paradigm has been found to be useful for programming reactive applications like avionics, process controllers and man-machine interfaces. Reactive programs execute forever, maintaining continuous interaction with their environment. The execution is a sequence of *reactions* in which environmental inputs are read and trigger output events. The characteristic feature of the synchronous paradigm is the *synchrony hypothesis* that the reaction time is zero so that outputs are produced simultaneously with the inputs. This is a simplifying assumption that makes the design and verification of such systems easier. It is also realistic since the CPU speed is an order of magnitude higher than I/O device speed.

Synchronous paradigm provides three novel constructs for building complex programs: synchronous parallelism, hierarchical composition and hiding. In a parallel program, the components run in lockstep with all of their reaction instants synchronized and interact with each other by means of broadcast signals. The hierarchy operator helps in structuring a complex programs as a hierarchy of functionalities; this is also useful for expressing preemption and watch-dog constructs useful for reactive systems. Hiding enables to abstract out internal details of systems. We define the refinement relation based on an abstract model of synchronous programs, called the the boolean I/O automata model (Maraninchi and Halbwachs, 1996).

The main results of the paper are:

(1) A concurrent program refines each its concurrent components.
(2) A hierarchical program refines its subprograms.

---

[1] Partially supported by the Indo-US project titled 'Programming Dynamical real-time systems'

(3) If a program satisfies a linear temporal logic formula, then any program that refines it also satisfies the same formula.

These three results are useful as they provide a syntactic method for verifying complex synchronous programs: in order to verify that a given program satisfies a formula, verify its syntactic subsystems for the same formula; the latter problem is usually simpler than the original problem. The basic idea of compositional verification is not new (Grumberg and Long, 1991), (Kurshan, 1994). The presented work can be seen as an extension of these results to the synchronous framework.

The organization of this abstract is as follows. Section 2 summarizes the synchronous paradigm and the semantic model. Section 3 describes the proposed refinement relation and shows the main results. Section 4 describes a verification scheme based upon these results.

2. SYNCHRONOUS MODEL OF REACTIVE PROGRAMS

Reactive programs, unlike transformational programs, execute forever, maintaining continuous interaction with their environment.Typically these programs find use in embedded applications like avionics, robotics and process control systems.

A programming model that has been found quite useful for reactive applications is the *synchronous model* (Berry and Gonthier, 1992), (Halbwachs, 1993), (Maraninchi, 1989), (Harel, 1987). This model makes a clear separation between the control and the data and I/O handling parts of a reactive system. The control part, called the *reactive kernel* is the most crucial and complex part. It has a high level signal interface through which it interacts with the I/O part of the system. The kernel is executed at fixed points of times called the reactive *instants*. At each instant, certain input signals are present and the kernel *reacts instantaneously* to these inputs by generating certain output signals. The reaction is instantaneous in the sense there is no delay in generating outputs in response to input events.

2.1 *I/O Boolean Automata Model*

A formal model of synchronous programs is the I/O boolean automata model due to (Maraninchi and Halbwachs, 1996). Let S be an arbitrary set of signal names and $\mathcal{B}(S)$ is the set of all boolean expressions over S. Also let $\mathcal{M}(S)$ be the set of all *complete monomials* which are simple conjunctions of signal names or their negations including all the signal names.

Definition 2.1. A boolean automata is a tuple (Q, q_0, I, O, T), where Q is a set of states, $q_0 \in Q$ is the initial state, I is a set of input signals, O is the set of output signals, and T is the transition relation given by $T \subseteq Q \times (\mathcal{B}(I) - \{false\}) \times 2^O \times Q$.

Boolean automata can be represented by edge labeled directed graphs as illustrated in Figure 1. Two properties that can be associated with boolean automata are *reactivity* and *determinism*: An automata is reactive if it reacts to all possible input situations. An automata is deterministic if in any state and for any input situation, there is at most one next state.

2.2 *Operations on Automata*

Synchronous languages provide three important constructs for programming the kernel: concurrency, hierarchy and hiding. Corresponding to these constructs, three operators are defined over automata:

Concurrency
Concurrency in synchronous languages is synchronous in the sense that concurrent components run in lock-step with all their reaction instants synchronized. Also the components interact with each other and with the environment by means of broadcast signals. Concurrency is modeled as

Definition 2.2. **Synchronous Product**: Given two boolean automata $A_i = (Q_i, q0_i, I_i, O_i, T_i)$, $i = 1, 2$, the synchronous product automata $A = A_1 \| A_2$ to be the automata $(Q_1 \times Q_2, q0_1 \times q0_2, I_1 \cup I_2, O_1 \cup O_2, T)$, where T is given by

$$T = \{((q_1, q_2), b_1 \wedge b_2, o'_1 \cup o'_2, (q'_1, q'_2)) | \\ (q_i, b_i, o'_i, q'_i) \in T_i, i = 1, 2, \neg(b_1 \wedge b_2)\}$$

The transitions of the synchronous product A are consistent joint transitions of the components.

Hierarchy
The hierarchical composition of two programs A_1, A_2, results in a program P that behaves like P_1 in all the states of P_1, except that in one of the states, it behaves *in addition*, like P_2. This is modeled by

Definition 2.3. **Hierarchical Product**: Let $A_i = (Q_i, q0_i, I_i, O_i, T_i), i = 1, 2$ be two automata. Then the hierarchical product automata $A = (A_1)_{(q\|A_2)}$ is $(Q, q0, I_1 \cup I_2, O_1 \cup O_2, T)$, where $q \in Q_1$ and

$Q = (Q_1 \setminus \{q\}) \cup (\{q\} \times Q_2)$
$q0 = q0_1$ if $q \neq q0_1$
　　$q0_1 \times q0_2$, otherwise

$$T = \{(q_1, b, o, q_2) \in T_1 | q_1, q_2 \neq q\} \cup$$
$$\{(q_1, b, o, (q, q0_2)) | (q_1, b, O, q) \in T_1\} \cup$$
$$\{((q, q_1), b_1 \wedge b_2, o, (q, q_2)) |$$
$$(q, b_1, \emptyset, q) \in T_1, (q_1, b_2, o, q_2) \in T_2,$$
$$\neg(b_1 \wedge b_2)\} \cup$$
$$\{((q, q_1), b_1 \wedge b_2, o, q') | (q, b_1, o_1, q') \in T_1,$$
$$(q_1, b_2, o_2, q_2) \in T_2,$$
$$o = o_1 \cup o_2, \neg(b_1 \wedge b_2)\}$$

The transitions of A, as given by the above four sets, are: (i) those of A_1 with the state q not being a source nor target state, (ii) those of A_1 in which the target state is q; after taking such a transition, the hierarchical automata enters the initial state of A_2, (iii) those of A_2 in which A_1 is idle and (iv) those in which q is exited.

Localization/Hiding

Using this construct, the influence of a signal can be localized and hidden to the outside world. This is modeled as follows:

Definition 2.4. Given a boolean automata, $A = (Q, q0, I, O, T)$, and a signal a, A^a is a new automaton $(Q, q0, I \setminus \{a\}, O \setminus \{a\}, T')$, where

$$T' = \{(q_1, b[a \mapsto true], o \setminus \{a\}, q_2) |$$
$$(q, b, o, q) \in T, b[a \mapsto true] \neq false, a \in o\} \cup$$
$$\{(q_1, b[a \mapsto false], o, q_2) |$$
$$(q, b, o, q) \in T, b[a \mapsto false] \neq false, a \notin o\}$$

The transitions in T' are derived from two kinds of transitions in T: (i) All those of T in which a is output and a is assumed to be present in the input; the latter is represented by substituting the truth value T for a in the input expression b and (ii) all those transitions in which a is not output by A and a is assumed to be absent in the input. Note that a does not occur neither in the input condition nor in the output set of A^a.

2.3 *Trace semantic model*

The boolean automata are descriptions of detailed behavior of reactive systems. For the purpose of verification, we require a more abstract model and hence propose a trace model. The trace model associates with each program the set of all its execution traces that record at each instant a condition on the input signals and a set of output signals that are generated. An execution trace or simply a trace is an infinite sequence of ordered pairs of input condition and output signals that are required/generated in a particular execution of the automaton. Let τ, σ with possible subscripts stand for traces. Given a trace τ, $\tau[i]$ is the ith element of the trace which is an ordered pair: the first element of the pair is a boolean formula over input signal names while the second component is a subset of output signal names; the input boolean formula is in complete monomial form.

Let $\tau[i].I, \tau[i].O$ denote the input condition and output signal set of the pair $\tau[i]$ respectively; also let $\tau.I, \tau.O$ stand for the sequences of input monomials and output sets (respectively) corresponding to the elements of τ. Further, in the rest of this paper, we lift the boolean operators and set operators to sequences $\tau.I, \tau.O$ for any trace τ. Given a boolean automaton $A = (Q, q0, I, O, T)$, the trace set associated with it is denoted by $\mathcal{T}(A)$ and given by

$$\mathcal{T}(A) = \{\tau | \exists (q_i, b_i, o_i, q_{i+1}) \in T, q_0 = q0,$$
$$\tau[i].I \to b_i, o_i = \tau[i].O, i/geq0\}$$

Note that the input condition of a transition in an automata need not be in complete monomial form and hence in the above definition, it is required that $\tau[i].I$ logically implies the condition b_i; thus there may be more than one monomial implying a boolean condition.

Operations on the trace sets corresponding to the boolean automata operations can also be defined, which are described in the full paper.

3. REFINEMENT OF SYNCHRONOUS PROGRAMS

We use the behavioral trace semantics of synchronous programs for defining the refinement relation. Let A and B be arbitrary boolean automata. Then,

Definition 3.1. A is a *refinement of* B (or A **refines** B), denoted by A **ref** B, provided

$$\forall \tau \in \mathcal{T}(A), \exists \sigma \in \mathcal{T}(B):$$
$$\forall i : (\tau[i].I \to \sigma[i].I) \wedge \tau[i].O \supseteq \sigma[i].O$$

Or simply,

$$\forall \tau \in \mathcal{T}(A), \exists \sigma \in \mathcal{T}(B):$$
$$(\tau.I \to \sigma.I) \wedge (\hat{\tau.O} \to \hat{\sigma.O})$$

where $\hat{\tau.O}$ is the sequence $b_0, b_1, \cdots$ with each b_i being the conjunction of the signals in the set $\tau[i].O$; note further that $\to$ is extended point wise to sequences in the above definition.

Intuitively, a refined system is one which may have less number of traces, may not have a trace for a given input sequence; also it may generate more outputs than the abstract version.

We have the following simple results:

Lemma 3.2. If A refines B, then $I_A \supseteq I_B$, where I_A, I_B are the sets of input signal names of A and B respectively.

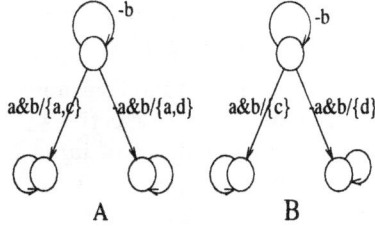

Fig. 1. Before hiding

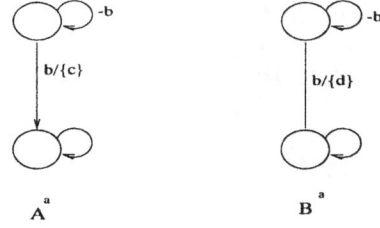

Fig. 2. After hiding

Lemma 3.3. ref is a pre order. It is a partial order over the set of all deterministic systems.

Reflexivity and transitivity are easy to prove. Antisymmetry requires that programs are deterministic: Suppose that A ref B and B ref A. Then $I_A = I_B$ follows from Lemma 3.2. Suppose τ a trace of A. Then there is a trace σ of B such that $\tau.I \to \sigma.I$ and $\tau.\hat{O} \to \sigma.\hat{O}$. Also since B ref A, there is $\tau' \in A$ corresponding to σ such that $\sigma.I \to \tau'.I$ and $\sigma.\hat{O} \to \tau'.\hat{O}$. Since, the input conditions in a trace are complete monomials over the same alphabet, it follows that $\tau.I = \sigma.I = \tau'.I$. Then determinacy of A, B implies that $\tau.O = \tau'.O$. That is $\tau = \tau' = \sigma$. Thus every trace of A is in $\mathcal{T}(B)$. Using a similar argument, we can prove that any trace of B is also in A and hence the result.

We have the following useful results:

Lemma 3.4. For any automaton A, B, $(A\|B)$ ref A and symmetrically $(A\|B)$ ref B. Further, for any state q in B, $(A_{(q\|B)})$ ref A.

We have the following interesting substitutivity results as well.

Lemma 3.5. If A ref B then $(A\|C)$ ref $(B\|C)$ for any C. Also $C_{(q\|A)}$ ref $C_{(q\|B)}$ for any C and q in C.

Our programme of modular verification would have been smoother had we had the positive result of substitutivity of ref under the localization operation also. But unfortunately we have,

Lemma 3.6. There exist deterministic and reactive A, B, A^a, B^a such that A ref B but A^a ref B^a is not true.

Proof: Consider A, B as given in Figure 1. It is easy to check that A, B are reactive and deterministic and also A ref B. Then A^a, B^a are given as in Figure 2. It is also easy to see that these are also reactive and deterministic but A^a ref B^a does not hold.

Now we define a stronger refinement relation sref that satisfies slightly weaker properties but which might serve our purpose.

Definition 3.7. Given A, B boolean automata, A is a **strong refinement of B** (or A **strongly refines** B) with respect to a signal a, denoted by A sref$_a$ B, provided

$$\forall \tau \in \mathcal{T}(A), \exists \sigma \in \mathcal{T}(B) :$$
$$(\tau.I \to \sigma.I) \wedge (\tau.\hat{O} \to \sigma.\hat{O}) \wedge$$
$$\forall i : (a \in \tau[i].O \to a \in \sigma[i].O)$$

sref is a stronger refinement relation:

Lemma 3.8. If A sref$_a$ B then A ref B.

Like ref, sref is, in general a pre order and partial order over deterministic systems.

Definition 3.9. An automaton A is said to be a *producer* of a signal, say a, if there exists a transition in A whose output part contains the signal a.

sref enjoys weaker compositional properties than ref as stated by the following results:

Lemma 3.10. For any automaton A, B, suppose that B is *not* a producer of a signal a, then $(A\|B)$ sref$_a$ A; symmetrically $(A\|B)$ sref$_a$ B if A is not a producer of a.

Lemma 3.11. For any pair of automata A, B and any state q in A, $(A_{(q\|B)})$ sref$_a$ A, provided B does not produce a.

This relation has the required congruence properties:

Lemma 3.12. If A sref$_a$ B then for any C, $(A\|C)$ ref$_a$ $(B\|C)$ and $C_{q\|A}$ ref$_a$ $C_{q\|B}$; Further A^a sref$_a$ B^a.

Proof of substitutivity of sref with respect to the parallel composition and hierarchical operation is easier. We prove the result only for the localization operation. That is we prove that whenever A sref$_a$ B then A^a sref$_a$ B^a.

To show that A^a ref B^a, consider any trace τ_A of A_a. There exists a trace σ_A of A such that for any $i \geq 1$

$$\tau_A[i].I = \sigma_A[i].I[a \mapsto true], \text{ if } a \in \sigma_A[i].O$$
$$\sigma_A[i].I[a \mapsto false], \text{ if } a \notin \sigma_A[i].O$$
$$\tau_A[i].O = \sigma_A[i].O \setminus \{a\}$$

Since A sref$_a$ B, there exists $\sigma_B \in \mathcal{T}(B)$ with $\sigma_A.I \to \sigma_B.I$ and $\sigma_A.O \to \sigma_B.O$. Corresponding to the trace σ_B, we can have a trace τ_B of B_a, i.e., the one that is related to the latter in the same way τ_A is related to σ_A; note that this is possible only because of the additional restriction imposed by sref. Now we prove that $\tau_A.I \to \tau_B.I$ and $\tau_A.O \to \tau_B.O$. The following cases have to be distinguished:

(i) $a \in \sigma_A[i].O, a \in \sigma_B[i].O$. Then

$\tau_A[i].I = \sigma_A[i].I[a \mapsto true]$
which implies
$\sigma_B[i].I[a \mapsto true] = \tau_B[i].I$ and
$\tau_A[i].O = \sigma_A[i].O \setminus \{a\} \supseteq \sigma_B[i].O \setminus \{a\} = \tau_B[i].O$

(ii) $a \notin \sigma_A[i].O, a \notin \sigma_B[i].O$. Then

$\tau_A[i].I = \sigma_A[i].I[a \mapsto false]$
which implies
$\sigma_B[i].I[a \mapsto false] = \tau_B[i].I$ and
$\tau_A[i].O = \sigma_A[i].O \supseteq \sigma_B[i].O = \tau_B[i].O$

Note that the other two cases are not possible since A sref$_a$ B. Hence the lemma.

4. A TEMPORAL LOGIC

In this section, we propose a version of propositional linear temporal logic suitable for specification of reactive systems. In this logic, the propositional symbols are from the signal set $\mathcal{S}$ itself; each symbol denoting its occurrence.

Atomic formulae The atomic formulae are of the form $f \hookrightarrow g$, where f, g are arbitrary propositional formulas over $\mathcal{S}$.

Formulae

(1) All atomic formulae and their boolean combinations are formulae.
(2) If ψ, ϕ are formulae then so are $\Box\psi, \Diamond\psi, \circ\psi, \phi \, \mathcal{U} \, \psi$ and their boolean combinations.

Semantics The semantics is standard (Manna and Pnueli, 1992) except for atomic formulae. Atomic formulae are interpreted over records of traces while temporal formulae over traces. Let b be an arbitrary boolean formula over signal names, o a finite subset of signal names and τ any arbitrary trace of a boolean automata.

$(b, o) \models f \hookrightarrow g$ if $b \to f$ and $\hat{o} \to g$

$\tau \models \phi$ if $(\tau, 1) \models \phi$, ϕ - any arbitrary formula.
$(\tau, j) \models \phi$ if $\tau[j] \models \phi$, ϕ - pure boolean formula.

We define a subset of formulae called *negation free formulae*. A negation free formula is equivalent to a formula in which no negation symbol appears except perhaps 'inside' atomic formulae.

We define now when an automaton can satisfy a temporal logic specification.

Definition 4.1. An automaton A satisfies a temporal logic specification ϕ, denoted by A sat ϕ provided that for any $\tau \in \mathcal{T}(A)$, we have that $\tau \models \phi$.

We have the following important result:

Lemma 4.2. Suppose that ϕ is a negation free formula. If B sat ϕ and A ref B then A sat ϕ.

Proof: Without loss of generality, we shall assume that ϕ does not contain any negation (except inside atomic formulae). Suppose that A ref B and B sat ϕ. We prove that A sat ϕ by induction on the structure of ϕ.

Base: ϕ is a formula of the form $f \hookrightarrow g$. Consider any trace τ of A. We have to prove that $\tau[1] \models \phi$. By assumption, there is $\sigma \in \mathcal{T}(B)$, such that $\tau[1].I \to \sigma[1].I$ and $\tau[1].O \supseteq \sigma[1].O$. The latter implies that $\tau[\hat{1}].O \to \sigma[\hat{1}].O$. Since B sat ϕ, we have that $\sigma[1].I \to f$ and $\sigma[\hat{1}].O \to g$. Then the required result follows from these two facts.

Induction Assume that the result holds for arbitrary formulae ϕ, ψ and prove the result for $\phi \wedge \psi, \Box\phi, \circ\phi$ and $\phi \, \mathcal{U} \, \psi$. We prove the result for the last case. The proof for the rest of the cases are similar.

Suppose that A ref B and B sat $\phi \, \mathcal{U} \, \psi$. Consider any trace τ_A of A. By assumption there exists τ_B of B such that $\tau_A.I \to \tau_B.I$ and $\tau_A.\hat{O} \to \tau_B.O$. Further there exists $j > 0$ such that

(1) $(\tau_B, j) \models \psi$ and
(2) for each $k : 1 \leq k < j, (\tau_B, k) \models \phi$.

It is easy to see that the above two facts hold for τ_A as well. Hence A sat $\phi \, \mathcal{U} \, \psi$.

5. VERIFICATION OF SYNCHRONOUS PROGRAMS

Summarizing the results of the previous sections, we have that properties of appropriate subcomponents in a large system holds in the whole system also. This gives rise to a simple compositional verification strategy: to verify a global property of a system, decompose the property into a number of local properties each of which holds in a subsystem. Verification of local properties would be simpler as subsystems would be smaller than the whole system; also they can proceed in parallel. These results are useful in compositional development of large systems and their properties: any property of a subsystem can be taken to be a property of the whole system being developed. This verification strategy is, in general, not automatic as neither the decomposition of a global property into local properties nor the identification of subsystems is easy.

Here we propose a strategy for automatic identification of subsystems based upon the specification. This is very simplistic but may be effective in large systems.

Given a formula f, let $out(f)$ denote the set of all output signals that appear in f. Further we need the following definitions:

Definition 5.1. An automaton A <u>may generate</u> a set of signals o' provided A has a transition with the label e/o such that $o \cap o' \neq \emptyset$.

Given a boolean formula e, let $pos(e)$ denotes the set of all signals which may be required to be true in order for e to be true. More precisely,
$pos(e) = \{S | (e \wedge S) \text{ is satisfiable }\}$.

Definition 5.2. Given a pair of automata A, B, we say that A <u>may trigger</u> B provided there is a transition in each of A, B having the labels e_1/o_1, e_2/o_2 such that $o_1 \cap pos(e_2) \neq \emptyset$.

Intuitively, A may trigger B provided there is a transition in A which when taken will generate a signal which may trigger a transition in B.

Definition 5.3. Two sub automata A, B of C are said to be concurrent denoted A **conc** B, provided that there is an execution of C in which control simultaneously in both A, B.

Now we give a sketch of an algorithm to compute the required sub automaton for a given formula.

Algorithm:

Input: GA (automaton), f (Temporal formula)
Output: GB, a sub-automaton of A.
Steps:

(1) $H := \emptyset$;
(2) For each C, a subcomponent of GA, if C may generate $out(f)$, then $H := H \cup \{C\}$.
(3) $H := H \cup \{C\}$ for each subcomponent C such that there is a subcomponent C' with C **conc** C' and C triggers C'.

(4) Repeat this until there is no change in H.
(5) GB is the subcomponent of GA containing only those components in H.

We have the following result:

Lemma 5.4. If GA satisfies f then GB also satisfies f.

The proofs of all the results and the illustration of the approach are given in the full paper.

6. REFERENCES

Back, R.J.R. (1985). Refinement of programs. *Acta Informatica*.

Berry, G. and G. Gonthier (1992). The Esterel synchronous programming language: Design, semantics, implementation. *Science Of Computer Programming* **19**(2), 87–152.

Grumberg, O. and D.E. Long (1991). Model checking and modular verification. In: *Proc. of Concur '91*. Springer verlag, LNCS 527.

Halbwachs, N. (1993). *Synchronous Programming of Reactive Systems*. Kluwer Academic Publishers. Dordrecht.

Harel, D. (1987). Statecharts: A visual formalism for complex systems. *Science of Computer Programming* **8**(3), 231–274.

Kurshan, R.P. (1994). *Computer-aided Verification of Coordinating Processes*. Princeton University Press.

Manna, Zohar and Amir Pnueli (1992). *The Temporal Logic of Reactive and Concurrent Systems Specification*. Springer-Verlag.

Maraninchi, F. (1989). Argonaut: graphical description, semantics and verification of reactive systems by using a process algebra. In: *Workshop on Automatic Verification Methods for Finite State Systems*. LNCS 407, Springer Verlag. Grenoble.

Maraninchi, F. and N. Halbwachs (1996). Compositional semantics of nondeterministic synchronous languages. In: *Proc. of ESOP '96*. LNCS Vol. 630, Spinger Verlag.

Morgan, C.C. (1990). *Programming from Specifications*. Prentice-Hall Int.. New-York.

Copyright © IFAC Distributed Computer Control Systems,
Sydney, Australia, 2000

Systems Modelling and Identification in CAN based Distributed Control Systems

Alexandre Manuel Mota, Pedro Fonseca, José A. Fonseca
{alex, pf, jaf}@det.ua.pt

DETUA - IEETA
Universidade de Aveiro
P-3810-193 Aveiro, Portugal

Abstract: Distributed control systems performance can be affected by the occurrence of jitter in the messages that carry relevant data such as sample and actuation variables. This jitter comes from the influence of messages from other sources and thus depends on factors such as the distribution of controller tasks and the medium access control used in the network. When a CAN- controller area network is considered, this jitter can be modelled as a random variable with a gamma distribution. In this paper a study of the influence of this specific type of jitter in system identification with recursive implementation is presented. The results are derived in an adverse situation when both sampling and actuation data suffer from jitter. It is shown that using a model that assumes fractional dead-time in the system leads to a much better parameter identification than when the problem is just ignored. *Copyright © 2000 IFAC*

Keywords: CAN Distributed Control Systems, Jitter, System Modelling, Identification

1. INTRODUCTION

Presently, distributed systems find wide dissemination in embedded control applications, particularly in real-time systems for the automotive and robotics fields. Most of them rely on a fieldbus [1] to interconnect a set of nodes. When periodic variables, such as the ones used in control applications, are to be transmitted, it is possible to impose an average transmission period but, due to the interaction of other periodic, sporadic or aperiodic traffic, it is rather difficult to obtain constant time intervals between successive instances of the same periodic variable. The variation of the variable's periods due to MAC (medium access control) is often called network-induced jitter and may have a negative impact in control loops [2]. Recently, [3] revisits the subject of the degradation of controller performance due to jitter in the sampled and in the actuation variables and [4] studies the problem when a CAN – Controller Area Network is concerned. The need for further research in communication jitter minimisation is also referred in [5]. This subject is also under investigation in the general real-time systems field, e.g. in [6] and in [7] where changes in periods of control tasks are considered.

In this paper the influence in system identification performance of a particular type of jitter induced by the specific MAC of CAN is discussed and some simulation results are presented. The problem is addressed from two angles: in one the messages carrying sensor/actuator data are considered the ones with higher priority; in the other load messages used to simulate charge in the fieldbus are given a higher priority than the previous ones.

2. IMPLEMENTING DISTRIBUTED CONTROL SYSTEMS

Distributed control systems follow most of the times a generic architecture such as the one depicted in figure 1. There, it is shown just one of the several control loops that may share the same communication infrastructure. This example of control loop includes five blocks: the controller, the sensor and the actuator, which are mandatory but may share the same node (if

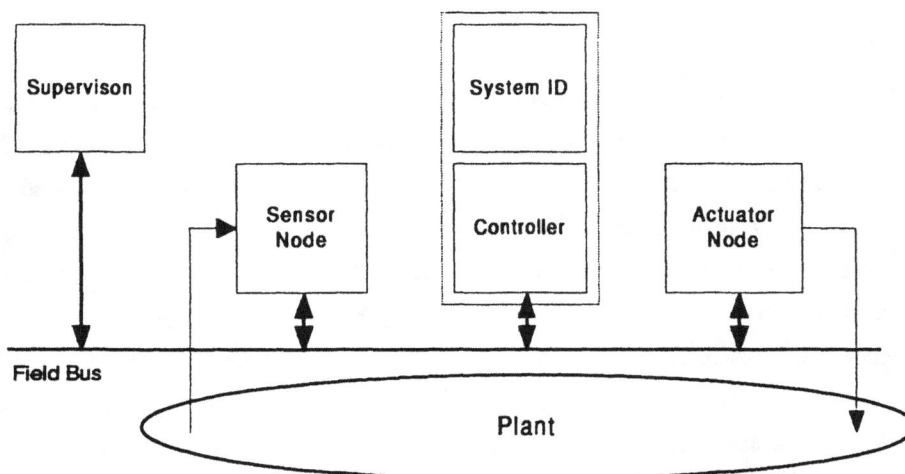

Figure 1 – Generic architecture of a distributed control system.

As the occurrence of jitter depends on the architecture of the distributed control system, this topic is briefly analysed in section 2. Here, it is also shown the results of the experiments carried on to obtain real data about the messages delay in CAN. This data is used in section 3 to model jitter in the sampling and in the controller output signals. The results presented in this section show that, if the presence of jitter is considered a priori instead of ignoring it, e.g. using a model that includes fractional dead-time, the system identification can result in a much better estimation of the system dynamic behaviour. Finally, in section 4, some conclusions are presented as well as ongoing research work following the same line.

all share the same node then the system becomes centralised) and the systems identification and the monitoring/supervision node which are optional.

It should be noticed that each of the nodes integrates a set of subsystems. For example the sensor node includes the sensor itself, an A/D interface and/or input port, a network interface and, almost always, a CPU, often of the microcontroller type. Other nodes such as the controller and actuator will also follow the same architecture. It is obvious that the supervision/monitoring node is most probably based on a PC or PLC and not in an embedded CPU. But in this case it will also operate with other control loops that share the same system.

From now on, the blocks of figure 1 are considered independent, as, for the purposes of this work, this is almost the worst case situation. In fact, this leads at least to the transmission of messages from the sensor to the controller, and from this one to the actuator. These messages will carry the periodic variables used in the controlling process. The system identification block shares a node with the controller, otherwise these two functions would also generate messages between then.

The production of the variables' values and the

unidentified messages in the message with the sensor data. The message is then received with an average periodicity that is, in principle, constant and equal to the initial period. However, the time interval between two consecutive instances of the message changes. This variation in the period is often called network-induced jitter.

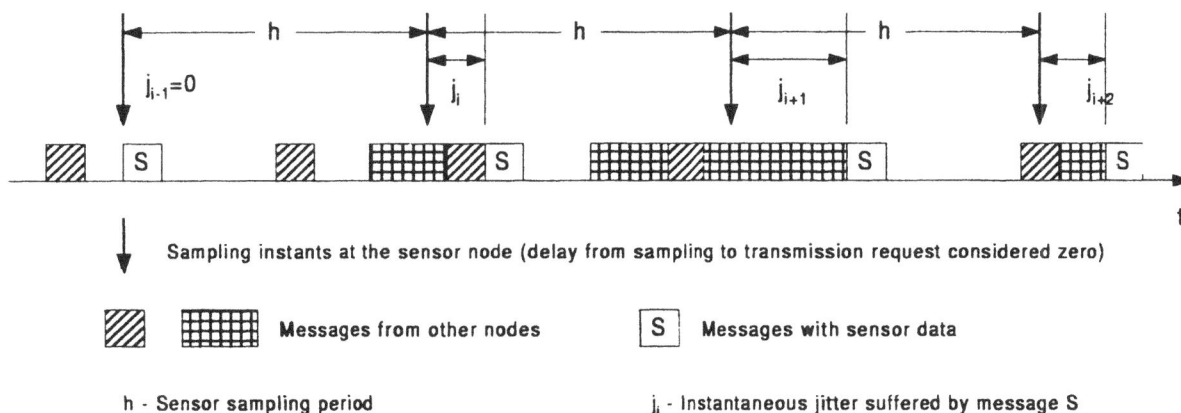

Figure 2 – Network-Induced jitter.

transmission of messages on the fieldbus are in general not synchronised. Most of the times, at the sensor node there is a timer generating the sampling instants and, when the data is obtained, the transmission services are called in order to transmit it. Tasks operating in other nodes that need this data, for instance the controller, must wait to its arrival in order to proceed executing. When they produce results, e.g. the actuation variable, these are requested to be transmitted as soon as they are produced. From the transmission point of view this type of operation is called event-triggered because transmission is requested in the sequence of an event not synchronised with the global system operation. Event-triggered operation is the most common approach in current distributed control systems.

The transmission of a periodic message carrying, e.g., sensor data acquired in specified instants with a constant sampling interval, may then be delayed by influence of a message being transmitted and of other messages waiting for the availability of the bus when these have higher priority from the MAC point of view. Figure 2 shows an example of the effect of

In a distributed system, factors such as the Medium Access Control used, the priority of the messages, the message scheduling algorithm (when the definition of the access to the medium is fixed before operation), initial phasing, affect the jitter value. In the case of CAN it is possible to obtain a measure of the jitter from a suitable representation of the messages delay as a random variable (r.v.). To do this, a set of experiments were conducted, consisting on sending the messages over a CAN network and measuring the messages delay. At the same time, some other messages were sent over the network, that corresponded to another application being executed (the load).

Figure 3 represents the experimental set-up used. It consists of two main groups of stations:

- 3 stations (1 Master and 2 Slaves) used to measure the message delay;

- 2 stations generating the network load (a set of periodic messages, with different sizes and periods).

A more detailed presentation of the experiments is described in [8].

The load message set is based on the PSA workbench [9]. The original set occupies 40% of the CAN load capacity at 125kbit/s. By varying the messages' periods, the resulting communication load can be changed. In this way, a disturbing load can be introduced on the communication channel, which could occupy from 5% to 95% of the network capacity.

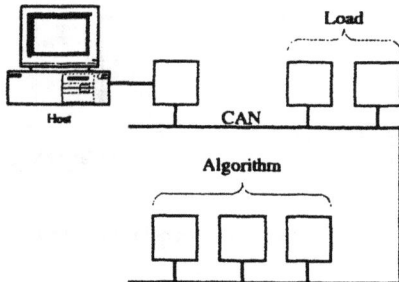

Figure 3 – Experimental set-up.

In these experiments, two different cases were considered, which were called "high" and "low". In the "high" case, the messages IDs are higher than load messages IDs (they have a lower priority). In the "low" case, the messages have lower IDs (higher priority).

In anticipation, one could foresee that two parameters would strongly influence the description of message delays: the message's IDs and the network load. Figures 4 to 7 present the histogram of message delays obtained experimentally. To these graphs, the p.d.f. of a Gamma r.v. with identical mean and variance were superposed.

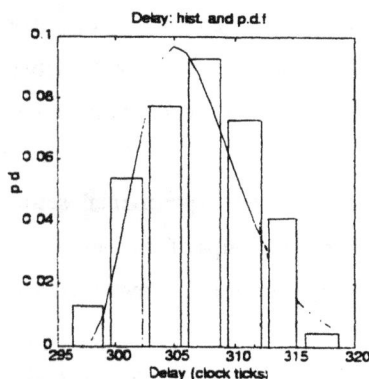

Figure 4 – Histogram & gamma pdf (Load=10%, low)

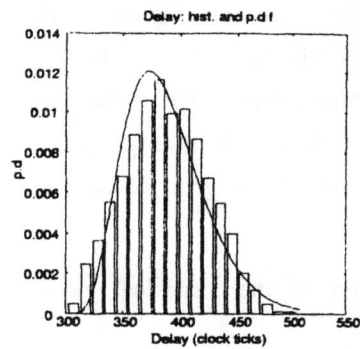

Figure 5 – Histogram & gamma pdf (Load=95%, low)

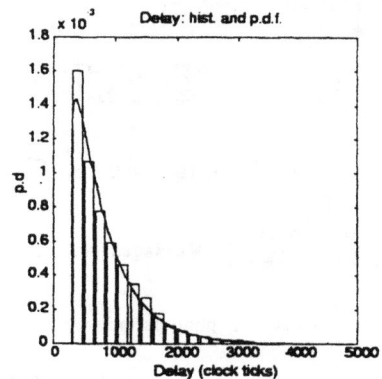

Figure 6 – Histogram & gamma pdf (Load=85%, high)

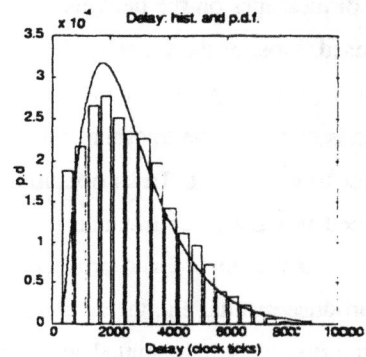

Figure 7 – Histogram & gamma pdf (Load=95%, high)

The results also show that the message delays in CAN can be represented by means of a Gamma distribution. Notice that this is valid under very different working conditions.

3. DEALING WITH JITTER IN SYSTEMS MODELLING AND IDENTIFICATION

The effect of jitter can be viewed as a perturbation that introduces a variable delay in the reading of the samples and in the actuation signal sent to the plant.

The existence of one or both of the situations depends on the distributed control system architecture. These situations are often named read-in and read-out jitter, respectively [3].

In figure 8 a block diagram of a system with read-out jitter is presented. There, it is shown that the discrete signal $u(k)$ is the input of a Zero-order-Hold (ZOH) which generates the continuous signal $u(t)$. This one suffers a delay, τ, before it is output to the plant. This delay is a random variable and thus, jitter will be present in the $u(t-\tau)$ signal.

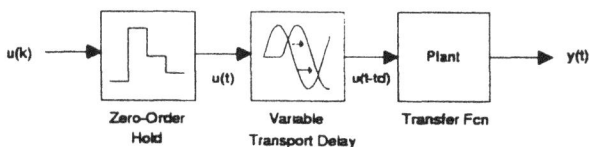

Figure 8 – Diagram of a system with read-out jitter.

Figure 9 now represents a block diagram of a system with read-in jitter. In this case the output of the system, $y(t)$, is delayed by an interval called τ, before the input of the ZOH. Again, this delay is a random variable and can then be considered jitter.

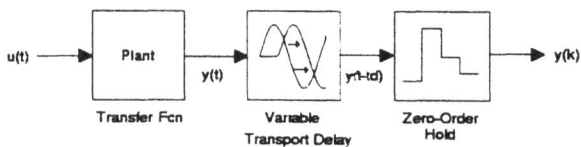

Figure 9 – Diagram of a system with read-in jitter.

Two situations can then be considered when one proceeds to the identification of a system with jitter. The first one is just ignoring it. The second is to try to take it in consideration in the system model. In this last case, a possible model for a SISO system could be:

$$\frac{dx(t)}{dt} = Ax(t) + Bu(t - \tau)$$
$$y(t) = x(t)$$

where τ is the variable delay which can obviously be considered as a dead time. As in most distributed systems it is possible to bound this dead time and if

this bound is the sampling period value ($h > \tau$), then the discrete model can be [10]:

$$y(kh + h) = \Phi y(kh) + \Gamma_0 u(kh) + \Gamma_1 u(kh - h)$$

where:

$$\Phi = e^{Ah}$$

$$\Gamma_0 = \int_0^{h-\tau} e^{As}\,ds\,B$$

$$\Gamma_1 = e^{A(h-\tau)}\int_0^{\tau} e^{As}\,ds\,B$$

The discrete transfer function is:

$$G(q) = \begin{bmatrix} 1 & 0 \end{bmatrix}(qI - \Phi)^{-1}\left(\Gamma_0 + \Gamma_1 q^{-1}\right)$$

This last equation represents the discrete model of a system with fractional dead time which has now a new zero that doesn't exist when τ is zero. It is also clear that Γ_0 and Γ_1 values are dependent on τ.

In order to study the validity of this model in CAN based distributed control systems, a simulation using four common control systems $\left(G(s) = \frac{1}{s+1}, \frac{1}{s}, \frac{1}{s^2} \text{ and } \frac{1}{s(s+1)}\right)$ was carried on. For each of them a recursive system identification procedure based on the least squares method with a fixed forgetting factor [11] was tested using either the simple model and the one with fractional dead-time. In the simulation the systems were affected by read-in and read-out jitter taken from the experiments described above.

Figures 10 and 11 show the results of the identification tests for all the referred systems, the first using the jitter values correspondent to the "low" designation and the second to the "high". There, it is shown the square residuals (estimation errors) average for the two type of models in function of the fieldbus load (from 5% to 95%). The * line corresponds to the more

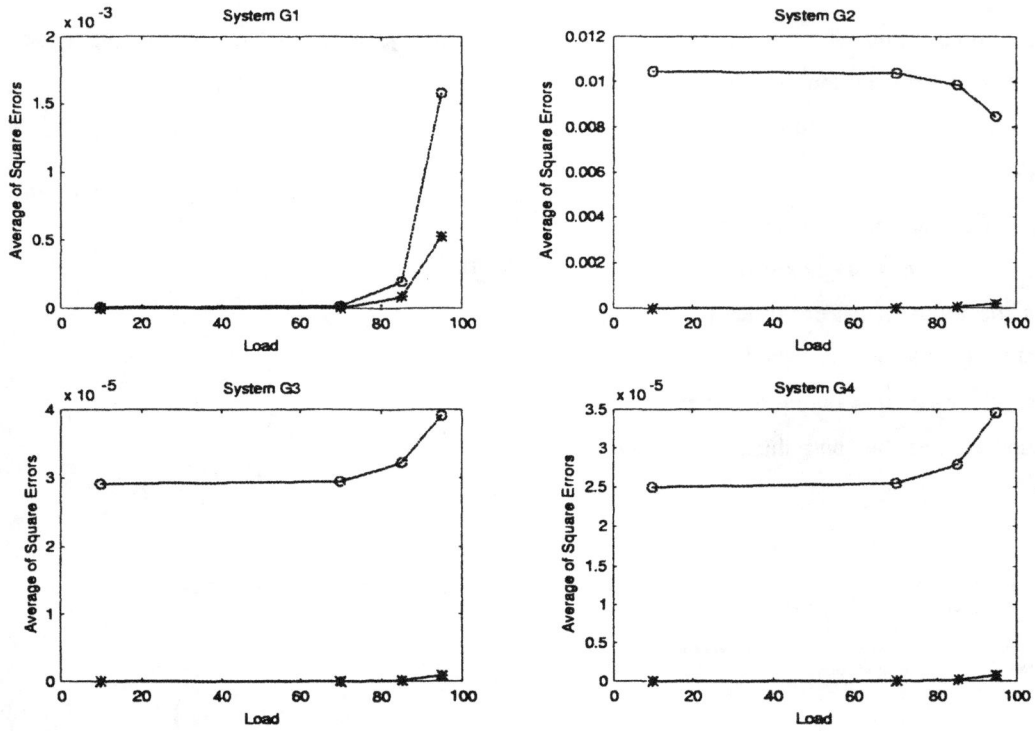

Figure 11 – Square Average of the Residuals versus the load (RLS-FF=0.98) – High Address

complex model (the one with fractional dead time) and the o line to the simple model.

It is clear from the figures that, when jitter is not taken into account, the model identification is poor when compared with the one that considers it as a fractional dead time. Values range from an improvement of 2.3

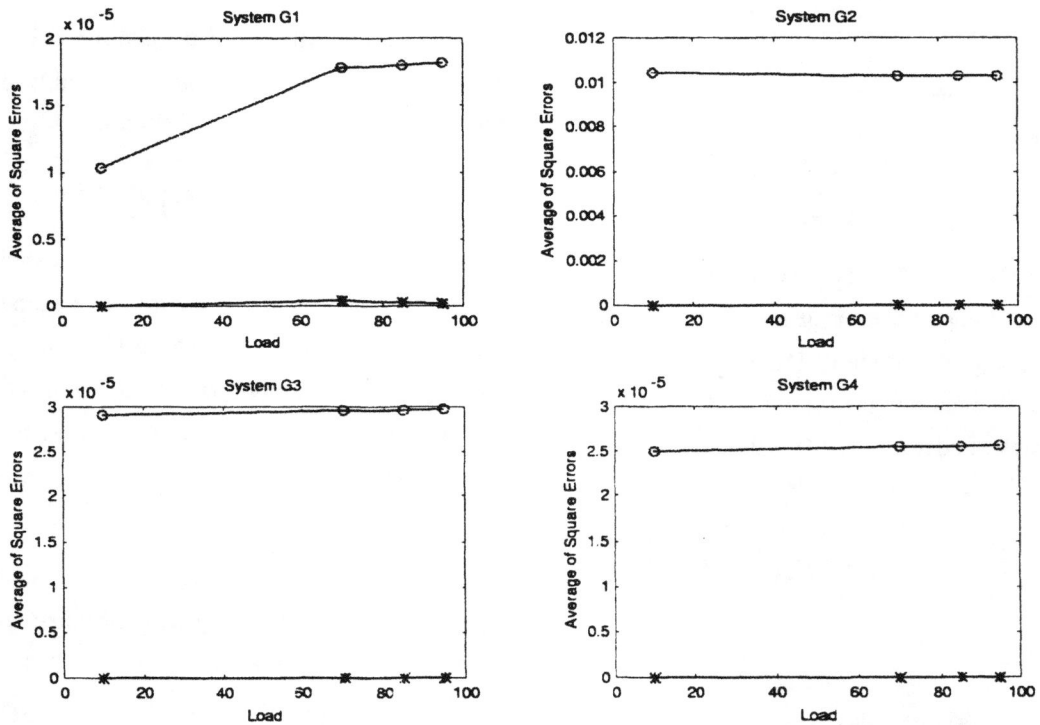

Figure 10 – Square Average of the Residuals versus the load (RLS-FF=0.98) – Low Address

88

times (G1 system with 85% load and low message priority – "high" case) to an improvement of 7700 times (G2 system with 70% load and low message priority – "high" case). Also, the more complex model shows in most situations some immunity to the influence of high transmission loads, independently of the priorities chosen for the relevant messages.

4. CONCLUSIONS AND FUTURE WORK

In this paper the problem of network induced jitter was presented and a preliminary study on its influence in systems identification was briefly discussed. The results presented come from the simulation of four simple and typical control systems subject to jitter at the sampling data and at the actuation variable levels. Jitter values come from a set of experiments carried on with a CAN network in which message delays were measured in different load conditions and with different message priorities. In the simulation results presented it was used a recursive identification algorithm based on least squares with a fixed forgetting factor. The results show that the use of a model considering a fractional dead-time lead to a much more accurate system identification than if jitter influence is ignored.

Some problems occurring from this work are still under investigation. One of them is the study of the influence of the forgetting factor in the identification performance under jitter conditions. Some specific techniques such as the directional forgetting factor [12] are also to be analysed in this situation. Also, it is should be pointed that further work must be done in order to evaluate the impact of the identification improvement in the closed loop behaviour of, for example, adaptive controllers.

REFERENCES

[1] Thomesse , "A Review of the Fieldbuses", Annual Reviews in Control, 22 pp. 35-45, 1998.

[2] Hong, S., "Scheduling Algorithm of Data Sampling Times in the Integrated Communication and Control Systems", IEEE Trans. Control Syst. Techn., Vol. 3, N° 2, June 1995.

[3] Stothert, etal "Effect of Timing Jitter on Distributed Computer Control System Performance", Proc. 15 IFAC Workshop DCCS'98 – Distrib. Comp. Control Syst., Sept. 1998.

[4] Juanole, "Modélis. Éval. Protocol MAC du Réseau CAN", École ETR'99, France, Sept. 1999.

[5] Decotignie, "Future Directions in Fieldbus Research and Development", Proc. FeT '99 - Fieldbus Syst. and Appl. Conf., Germ., Sept. 1999.

[6] Cervin, "Improved Scheduling of Control Tasks", Proc. 11th Euromicro Conf. Real Time Systems, June 1999.

[7] Shin, etal "Adaptation and Graceful Degradation of Control System Performance by task Reallocation and Period Adjustment", Proc. 11th Eurom. Conf. Real Time Syst., June 1999.

[8] Fonseca, P. – Modélisation et validation des algorithmes non-déterministes de syncronisation des horloges", Phd Thesis, Universidade de Aveiro, 1999.

[9] N. Navet and Y.-Q. Song, "Design of a Reliable Real-Time Distributed Applications over CAN", *INCOM'98 - IFAC Int. Symp. on Information Control on Manufacturing*, Nancy/Metz, France, 1998.

[10] Karl Astrom & Bjorn Wittenmark, "Computer Controlled Systems: Theory and Design", Prentice-Hall, 1990.

[11] Lennart Ljung, "System Identification - Theory for the User", Prentice-Hall, 1987.

[12] Tore Hägglund, "New Estimation Techniques for Adaptive Control", Phd Thesis, Lund Institute of Technology, 1983.

Copyright © IFAC Distributed Computer Control Systems,
Sydney, Australia, 2000

A CONTROL THEORY APPROACH FOR CONGESTION CONTROL IN INTRANETWORK

Priyadarsi Nanda

Network Research Lab,
School of Computer Science and Engineering,
University of New south wales, Sydney, Australia
(priyan@cse.unsw.edu.au

Abstract: This paper presents a congestion control technique in an intranetwork gateway by using a control theory approach. Such an approach is very much necessary to determine how essentially information regarding congestion can be sent to the hosts as a precautionary measure in time. As a result those hosts connected in the intra net will take necessary steps to avoid further worst case congestion in the gateway of the network. Our proposed scheme can be viewed as a two tier model, where admittance of various connections is done according to the traffic parameter declaration by the hosts in the first step and once accepted, periodic signals can be sent to the hosts to regulate their flow of traffic in accordance to the declared parameters. The main idea in our approach is to observe network states at different point of time at the gateway before sending any information to the hosts inside the network. In order to observe these states at the gateway, two important parameters such as process estimation and process parameters need to be calculated and to get them correctly, we propose to use Kalman filter in our approach. Kalman filter is derived from discrete time space analysis and is different from other filters in use, which estimates a process by using a form of feed back control. We believe, such an approach can be widely used in today's Internet technology making use of Resource Reservation Protocol (RSVP) and other signalling protocols to achieve a desired Quality of Service (QoS) amongst a variety of users. *Copyright © 2000 IFAC*

Keywords: QoS, Intranet, Congestion control, ABR service, Kalman filter

1.INTRODUCTION

Traffic control and resource management over the Internet is quite complex due to a mixture of different traffic types, as each type has a different Quality of Services (QoS) requirement. In support of such traffic types, networks must be designed properly to avoid worst case congestion scenario. It has attracted much attention due to the potential for improved utilisation through statistical multiplexing of traffic sources by using ATM switches in the Backbone network . For this reason we have considered to use ATM switch at the gateway of the intranets and these switches are connected through links making use of Available bit rate (ABR) category of services to the users. As ABR category of the service provides a variable bandwidth to the users, users can be informed about the available bandwidth through the use of certain control mechanism at the gateway. Inspite of several control algorithms being used by different researchers [6,7] to avoid congestion completely, there is always a trade off lies

between complexity of the algorithm and it's implementation. Current approaches are usually suited to a particular type of traffic, which can be broadly classified into two-category [8].

  a. Where focus is on restricting the traffic entering the network based solely on the source type and declarations, regardless of the current level of traffic generated by the source.

  b. Where algorithms include some kind of dependence on the current state of the network.

First approach focus on source restrictions and design for worst case. Though more flexible, focusing high utilisation, the second category doesn't provide guarantees for throughput or QoS.

Also these two approaches are quite disjoint in the sense that they focus on different traffic sources and the key elements of the algorithms have no common protocol or database structure. This paper discusses the above mentioned issues in the light of a novel proposed control algorithm using Kalman Filter. Though use of Kalman filter has been seen earlier [8] with the ATM networks, our approach takes a different turn applying it at the gateway of the intranet where the users can still make use of Internet protocols such as Transmission Control Protocol (TCP) and Internet Protocol (IP). Kalman filter is a very general filtering technique that can be applied for solution to such problems as optimal estimation, prediction, noise filtering and stochastic optimal control. Such filters can easily be programmed. Kalman filter is a set of mathematical equations providing an efficient computational solution of the least-square method. Such a technique will be quite useful in the context of communication networks as thousands of connections are accepted and released per second and each connection requires a proper QoS to be maintained for its flow of traffic.

This paper proposes a control mechanism to be executed at the gateway of the network. We propose an approach for traffic control mainly focussing on issues such as admission control and congestion avoidance through the use of Kalman filter. Before proposing our approach, we have discussed various works performed by different researchers in the past on congestion control making use of control theory. Section 3 of the paper proposes the framework of our approach with an emphasis on the use of Kalman filter. Finally we discuss our future works with the current structure of the Internet in section 4 and conclusion in section 5.

2. RELATED WORK

Application of control theory to the communication network has been studied by various researchers in the past [6,7,8]. The mechanism proposed in [6], discusses a classical feedback control system implemented through a controller at the gateway of the Intranet. The controller uses feedback rate signals from the access gateway to the Intranet hosts to control the traffic rates so as to keep the buffer in the gateway at a desired point. Information regarding optimum aggregate rate for the entire intranet with respect to current buffer level and available band width in the access link are periodically fed back to the controller by the gateway. How ever our proposed scheme never looks into the buffer level to calculate the desired signal but depends on the declared parameters by the hosts, and calculates the desired bandwidth for a requested connection. Depending on the

calculated bandwidth, if satisfied, the connection is accepted else an error signal will be communicated to the hosts to reduce the rate for the connection to be accepted. However, the scheme proposed in [7], makes use of control theory for regulating the flow into the network by implementing a kalman state estimator first and then changing the approach into a novel estimation scheme through the use of fuzzy logic. But this approach did not consider any explicit rate feedback from the gateway to the sources. As a result, system stability was the major component and was not suitable for large Intranets. The scheme proposed in [8], deals with aggregate bandwidth scheme with the help of a unified framework for both controllable and uncontrollable traffic sources in an ATM network. The framework is based on a linear kalman filter, which provides a required QoS guarantee by reserving bandwidth for possible error estimation. Our proposed scheme is also based on a similar approach but considers Internet protocol (IP) as the major protocol without any modification to the structure of the protocol. Also we argue that, with the current structure of the Internet, involving integrated and differentiated services network, our control theory approach can be used very well for a desired QoS.

Next section of the paper discusses our proposed scheme. First we show how kalman filter can be used as a major tool in our framework and then propose the control structure of our framework.

3. PROPOSED CONTROL STRUCTURE

The control structure in our proposed architecture is based on kalman filter, which provides a recursive solution to the discrete-data linear filtering problem. The reason for such an approach lies in the fact that, kalman filter is a very general filtering technique that can be applied for solution to such problems as optimal estimation, prediction, noise filtering and stochastic optimal control. Such filters can easily be programmed. Kalman filter can be applied to both stationary and non-stationary process and can include initial conditions for the problems stated above[5]. Figure 1 below gives the proposed framework of our scheme and the control process implements a kalman filter for process estimation and parameter estimation. In this process, kalman filter tries to estimate the state $x \in R^n$ of a discrete-time controlled process that is governed by linear stochastic difference equation [9].

$$X_{k+1} = A_k x_k + B u_k + w_k \longrightarrow (1)$$

With a measurement $z \in R^m$ that is
$$Z_k = H_k x_k + v_k \longrightarrow \quad (2)$$
Here w_k and v_k are the random variables and represents the process and measurement noise. A is an n X n matrix and relates the state at time step k to the state at k + 1. B is an n X 1 matrix and relates the control input $u \in R^1$ to the state x. H is an m X n matrix and relates the state to the measurement z_k.

Two types of error, pre and post errors are important here to predict the future estimation and which can be stated as:

$$\text{Pre error} = e_k^1 \equiv x_k - \overline{x_k}'$$

$$\text{Post error} = e_k \equiv x_k - x_k'$$

With the above mentioned two fundamental equations and pre and post errors , the estimation process can derive a new state for which a new bandwidth can be estimated.

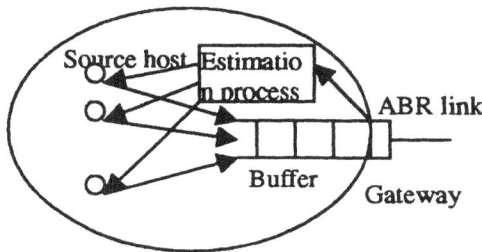

Figure 1: proposed framework

Our framework is based on the parameters declared by the sources for a connection establishment. Also new connections can be accepted along with old ones without any compromise of QoS. As shown above, sources are sending traffics through the gateway in the edge of the network. The gateways are connected to each other through ABR link. The estimation process in side the network is the main component of our proposed framework which includes a Kalman filter for process estimation and parameter estimation thereby considering the available band width in the ABR link, requested for a connection thereby generating a rate control signal to the sources. When no connection request is made by the sources and there are old connections still in process, it may so happen that, the available band width in the ABR link may be reduced due to certain congestion and the congestion notification is given back to the error estimation process by the gateway which is the first to encounter such situation. Then, the error estimation process computes the necessary reduction in source rate and sends

the sources necessary information through rate notification messages. The necessary algorithm for the framework needs certain careful consideration due to the fact that, inefficient error estimation may result in over or under utilisation of the link capacity in the output of the gateway.

However the complexity of the framework can be derived from the following four-principle [8] and the best one can be used to model our proposed structure.

a. The algorithm should operate explicitly in terms of the bandwidth so the cooperation between bandwidth management algorithms (BMA) on the connection and higher layer is straightforward.

b. used in an optimal way, ie.. the algorithm should use explicit measurements of the relevant network state and source traffic parameter declarations.

c. The algorithm should not be restrictive so that an unpredicted traffic increase should not be rejected if bandwidth is available.

d. There must be a single database structure supporting different traffic types and similar signalling protocols so as to reduce the complexity and cost of traffic control and bandwidth management.

Throughout our discussion, we consider to use available bit rate (ABR) category of traffic for carrying traffic streams over ATM backbones outside the Intranet.

The proposed structure for process estimation and parameter estimation can be implemented easily in the present scenario of Integrated services network (int-serv) and differentiated services network (diff-serv) with the help of Resource reservation protocol (RSVP). RSVP is a signalling protocol, which is used, in an int-serv network for resource reservation. RSVP provides receiver initiated setup of resource reservations for multicast or unicast data flows with good scaling and robustness properties. With the int-serv network, Sources can send traffic parameters with the help of T-Spec for establishing a connection to the destination network. The gateway/routers in the path can check for any available resources according to the specification mentioned in the T-spec. If resources are available, then the receiver can initiate a signal reserving the resources through the routers in the path and signals the source. More on RSVP can be found in [10]. The only difference between an int-serv and diff-serv is that, int-serv networks works on per flow of the traffic stream while diff-serv network works on aggregated flow. In both these two types of networks, resource management in terms of bandwidth reservation is quite essential and hence our proposed scheme can very well be applied to such situations. Finally we conclude our work with a note on future works to be performed in this area.

4. CONCLUSION

There are several control theoretics approaches for congestion control in an intra network particularly implemented at the boundary of the network to avoid congestion. Our approach gives a thought about how the control algorithm can estimate both process and parameters efficiently and depending on the error calculated can send necessary control messages to the sources either to increase or decrease the rate. This will help in avoiding further congestion in the network and the system can operate without any danger of failure. However, future work is required in implementing the algorithm and running simulation with a suitable set of parameters, resulting a proper control at the gateway of the Intranet. Finally with the current structure of the Internet where major research is done focussing more on the service classification (differentiated services network), we hope kalman filter can be used as a tool in resource allocation both in the intra-domain and inter-domain networks.

REFERENCES

[1] D.Bertsekas and R. Gallager, Data Networks, Prentice Hall, 3rd Edi.

[2] J.W. Roberts, Performance Evaluation and Design of Multiservice Networks, Commission of the European Communities, Cost 224 Project, 1992

[3] R.Jain, Congestion control and traffic Management in ATM Networks: Recent advances and a survey, Computer Networks and ISDN systems, 28, 1996, pp 1723 – 1738

[4] H. Gilbert et al, developing a cohesive traffic management strategy for ATM networks, IEEE Communication Magazine, Oct. 1991, pp 36-45

[5] K. Ogata, Discrete-time control systems, PrenticeHall, 1987

[6] M. Hassan, H. Sirisena, M. Atiquzzaman, A congestion control mechanism for enterprise network traffic over asynchronous transfer mode networks, Computer Communications 22 (1999), pp 1296-1306

[7] S.Keshav,A control-Theoretics Approach to flow control, ACM Comp. Communication. Rev. 21 (4) (1991) 3 - 15.

[8] Zbigniew Dziong, Marek Juda, and Lorne G. Mason, A frame work for Bandwidth management in ATM Networks , IEEE/ACM Transactions on Networking, vol.5, No. 1, February 1997

[9] Greg Welch and Gary Bishop, An Introduction to the Kalman Filter, TR 95-041, UNC-Chapel Hill, June6, 2000

[10] R. Braden, L.Zhang, S. Berson, S. Herzog, S. Jamin, Resource Reservation Protocol (RSVP), Internet draft, RFC 2205 September 1997

Copyright © IFAC Distributed Computer Control Systems,
Sydney, Australia, 2000

IMPROVING THE RESPONSIVENESS OF THE SYNCHRONOUS MESSAGING SYSTEM IN FTT-CAN

Paulo Pedreiras, Luís Almeida, José A. Fonseca

pedreiras@alunos.det.ua.pt,{lda,jaf}@det.ua.pt

DET-IEETA
Universidade de Aveiro
P-3810-193 Aveiro, Portugal

Abstract: A flexible distributed real-time communication system must support modifications to the message set which it conveys. These changes can require a degree of responsiveness ranging from a few milliseconds to some seconds.

In FTT-CAN protocol, the responsiveness of the synchronous communication system depends on the plan duration, which, in general cannot be set arbitrarily short. This paper presents a method that uses the asynchronous messaging system to temporarily convey the synchronous messages until the synchronous messaging system can handle them. Furthermore, methods to evaluate offline if a set of requests for modifications can be timely handled are presented. *Copyright © 2000 IFAC*

Keywords: Real-time communication, Real-time systems, Communication protocols, Distributed computer control systems, Fieldbus.

1 – INTRODUCTION.

1.1 – Levels of system responsiveness.

During normal operation, processes controlled by real-time computer systems experience phases of continuity as well as of changes (Fohler, 1995). Changes in the environment can be reflected in the real-time system as modifications to the task set, as well as to the message set when the system is distributed. Kopetz (Kopetz, 1997) states that resource utilization is improved if only those tasks that are needed in a particular operational mode are scheduled. In these circumstances the message set can change too. Consequently, a flexible real-time communication system must support changes to the message set which it conveys, namely allowing dynamic creation and elimination of message streams and change of parameters of existing ones. However, in the context of real-time systems, the timeliness of the communication system must always be guaranteed, even while changes to the message set are made. Thus, the requests for changes must be supported in a way that new

requirements are handled within adequate response time and without disturbing the timeliness of the remaining message streams.

The maximum time allowed between a change in the environment and the respective reaction in the control system is a critical parameter, which depends on the dynamics of both environment and control system. For example, consider a car traction control system in which a central unit receives information from wheels speed sensors and actuates on the breaking system if it detects that one or more wheels are losing grip. This kind of system can be implemented in a distributed fashion and, to improve resource utilisation, the wheels speed sampling rate might vary according to the driving conditions. When driving in a road with good adherence, the sampling rate can be lower. If the car suddenly enters a slippery road, the traction control system faces a sudden change in its operational conditions, requiring, among other things, a higher sampling rate. Since a car running at 100 Km/h travels 27,7m in a second, if the communication system requires 100ms to adjust the message set properties related to the sampling rate of the wheel sensors, the car travels about three meters until the system behaves accordingly to the new environmental conditions, jeopardizing the security of the driver and, eventually, other people. In this system a responsiveness of a few milliseconds is required. However, when (hopefully) the car returns onto good road again, the sampling rate can be reduced.

1.2 – About this paper.

The FTT-CAN (Flexible Time-Triggered communication on CAN) protocol is well suited to support the kind of system described above. Particularly, its synchronous message system, based on the time-triggered paradigm, can efficiently convey the message streams resulting from the periodic sampling of the wheels speed sensors.

This paper focuses on the responsiveness of the synchronous message system of FTT-CAN. It will be shown that some protocol key parameters cannot be adjusted only in function of the required responsiveness since they have wider implications. A method to improve the responsiveness to changes made to the synchronous message set is presented, and its implications in the protocol architecture are analysed. In section 2 the FTT-CAN synchronous and asynchronous messaging systems are briefly presented. The new method used to improve the responsiveness to changes in the communication requirements is presented in section 3. Section 4 presents methods to evaluate at pre-runtime if a given set of change requests can be timely handled. Finally, section 5 concludes the paper.

2. FTT-CAN BRIEF PRESENTATION.

The FTT-CAN protocol has been briefly presented in (Almeida *et al.*, 1999) and further developed in (Almeida, 1999). A feature that distinguishes this protocol from other proposals concerning time-triggered communication on CAN (Peraldi *et al.*, 1995) is that it supports dynamic communication requirements by using centralized scheduling with on-line admission control whilst the communication overhead is kept low by using the native distributed arbitration of CAN.

A Synchronous Requirements Table (SRTable) holds the properties of the synchronous message streams, namely: identifier, period, relative deadline, initial phase, maximum transmission time and priority. Using this information, the scheduler builds static schedules for consecutive fixed duration periods of time called plans. The creation of a plan is concurrent with the dispatching of the previous one.

As usual in table-based scheduling, a finite time resolution in used to express all the properties of the message set. This basic time unit is called Elementary Cycle (EC). The EC duration is fixed and set at pre-run-time.

Within each EC, the protocol supports two types of traffic, synchronous and asynchronous. The former one is time-triggered and its temporal properties (i.e. period, deadline and relative phasing) are represented as integer multiples of the EC duration. A particular node (Master), scans the current plan and generates a periodic message used to synchronize all other nodes in the network. The transmission

of this message represents the start of one elementary cycle (EC) and is known as EC trigger message (TM).

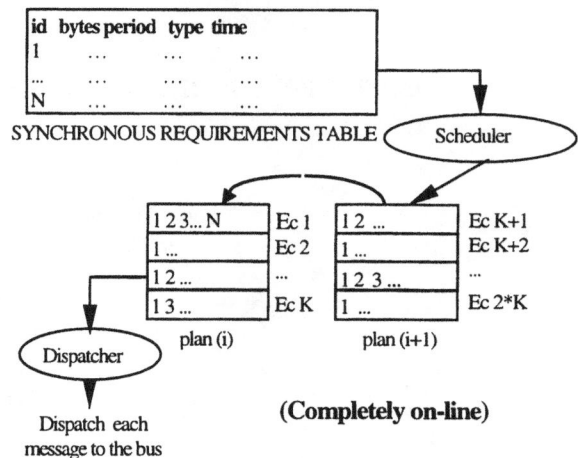

Fig. 1. The planning scheduler

The EC trigger message conveys in its data field the identification of the synchronous messages that must be transmitted by the producer nodes in that EC. The nodes that identify themselves as producers by scanning a local table containing the messages to be produced / consumed, transmit the respective synchronous messages in the synchronous phase of that EC (fig. 2). Collisions on bus access are resolved by the native distributed MAC protocol of CAN. This is known as the synchronous messaging system (SMS).

The FTT-CAN protocol also supports asynchronous traffic for event-triggered communication, with external control. This sort of traffic is transmitted during the periods of the EC not used by the synchronous messages. However, depending on how the desired temporal isolation between these two sorts of traffic is enforced, the asynchronous messaging system (AMS) can operate in one of two modes. In controlled mode any asynchronous message is transmitted only if it is guaranteed not to interfere with the timeliness of the EC trigger message or of the synchronous messages. This way, the temporal isolation between the synchronous and asynchronous traffic is strictly guaranteed. In uncontrolled mode, stations wishing to transmit asynchronous messages can try to do it as soon as they receive the respective requests from the application layer.

Although these messages may now cause a certain blocking to the transmission of synchronous ones, such blocking can be upper bounded by using a proper choice of identifiers. Thus, a bounded level of interference between the

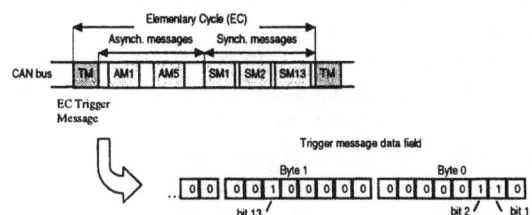

Fig. 2. EC Trigger Message data contents.

96

synchronous and asynchronous traffic is allowed, in exchange of a higher flexibility. In this case legacy nodes not committed with the FTT-CAN protocol are supported, if the identifiers are properly selected.

3 – IMPROVING FTT-CAN RESPONSIVENESS.

3.1 – Flexibility limits.

Once a change request is made concerning the current synchronous message set, a certain period of time elapses until that request takes effect at the bus level. This is referred to as the synchronous transient response time (STRT). Note that, when using SMS alone, the scheduler must, first, build a plan using the new requirements.

The STRT can be decomposed in three parts (see fig. 3): the time from the request to the end of the current plan, the plan in which the scheduler starts to take into account the new requirements, and finally the initial phase (φ) of the message stream relative to the beginning of the plan where changes are already reflected. The minimum value (marker A in fig. 3) occurs when cumulatively the request is made just before the end of one plan, and φ is zero. The maximum value occurs if the request is issued just after the beginning of one plan (marker B in fig. 3), and the initial phase has its maximum value. Therefore, the absolute bound for the synchronous transient response time, when using the SMS alone, varies between one and two plans plus the initial phase (as defined above) ($LPlan<STRT_{SMS}<2*Lplan+\varphi$). Since the $STRT_{SMS}$ is a direct function of the plan duration, the responsiveness can be improved by shortening the plan. However, the reduction of the plan duration increases the CPU load (Almeida *et al.*,1999)(Almeida,1999). Below a given value, the scheduler might not have enough time to build next plan in time, that is, before the dispatcher processes the current one. Moreover, some interesting properties of the planning scheduler, like the look-ahead feature (Almeida, 1999), are negatively affected by the reduction of the plan length. As a consequence, there is a lower bound to the plan duration, limiting the responsiveness that can be achieved this way. Another way to improve the responsiveness while still using the SMS alone is to start the scheduler as late as possible. Since the worst case execution time of the scheduler (wcetSch) can be estimated on-line (Almeida, 1999), using this approach the synchronous transient response time can be bounded to the interval:

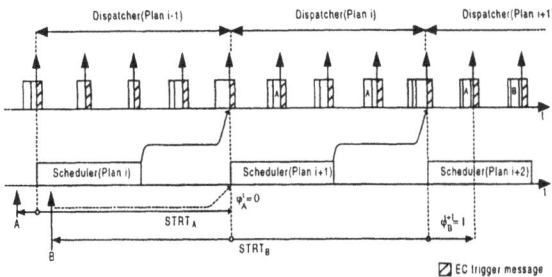

Fig. 3. SMS Responsiveness bounds.

$wcetSch < STRT_{SMS} < LPlan+wcetSch+o$, where LPlan stands for the plan duration.

3.2 – Improving FTT-CAN responsiveness.

As seen above, the responsiveness of the SMS when considered alone is upper bounded by the plan duration plus the scheduler execution time. Since these cannot be made arbitrarily short, further improvement to the responsiveness of SMS in FTT-CAN requires that change requests are handled even during the current plan, bypassing the planning scheduler for a short period of time, but without disturbing the other synchronous messages already scheduled. The proposed way of achieving this, consists in using the asynchronous messaging system (AMS) to produce the required message(s) until the requested changes are handled by the SMS as described in the previous section. This is shown in figure 4. Notice that the message associated with the change request (e.g. a new message stream) begins to be transmitted using the asynchronous message area. After the dispatcher starts processing the plan in which the new message parameters are reflected (plan i in figure 4), the system resumes normal operation, that is, as the message is included in the synchronous message area it is removed from the asynchronous one. The period of time during which the AMS is used to support the transmission of synchronous messages is referred to as synchronous support period (SSP). The Master station, by means of a specific control message, establishes the beginning and duration of the SSP for each change request. This message has higher priority than regular synchronous and asynchronous messages.

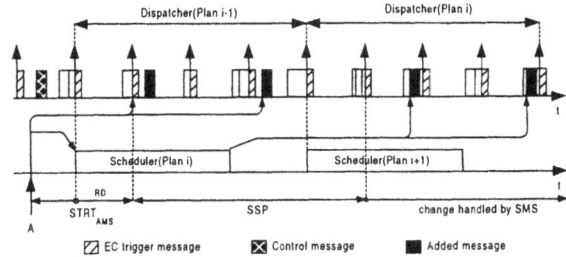

Fig. 4. Using the AMS to temporarily convey a new synchronous message.

The following relationship can be established between the STRT with and without the AMS support:

$$STRT_{AMS} = STRT_{SMS} - SSP \qquad (1)$$

If the change to the message set consists only in the addition of a new message, the process described above is sufficient. However, if the change request is performed over a message stream already present in the SRTable (e.g., to change the stream's period), the existing instances of the message in the SMS during the synchronous support period (SSP) should be eliminated. Those instances still use to the older parameters (before the change) while the updated instances are transmitted in the asynchronous area. Thus, the elimination is required to avoid replication of message production in both synchronous and asynchronous systems.

The elimination is achieved by filtering the trigger message. This filter is applied to the trigger messages just before they are sent. To eliminate one stream present in one plan already built, it is only necessary to set a bit in the filter.

Depending on the type of the change request that is made, one or several of the following actions can result:

1. A change of one bit in the filter;

2. The production of a control message to signal the start and duration of the SSP (synchronous support period);

3. A set of data messages produced in the AMS, during the SSP.

If the change request consists in the elimination of one message stream, only action 1 is required. However, if the change request consists in adding a new message, control and data messages will be produced in the AMS during the SSP (actions 2 and 3). If the change request concerns a modification in the parameters of an existing message (ex. period), actions 1,2 and 3 are required.

3.3 – Implementation issues.

From the operational point of view, several steps must be performed in order to process the request for a change to the synchronous message set. In figure 5 a flowchart describing the operational diagram of the proposed method for improving the responsiveness of the planning scheduler is presented.

When a change request to the synchronous message set is made, a schedulability test must be performed in order to filter out changes that would result in a non-schedulable message set. However, for the purpose of this work, we will consider that any requested change has already been analysed and it does not compromise the message set schedulability. In case the on-line analysis is performed, its execution time must be included in the STRT. Current work is being carried out in order to reduce such execution time (e.g. by using simple schedulability tests) so that its impact on the response time is minimized.

When a change request is accepted, the change is made to the SRTable, and then it is evaluated whether the response time requirements, expressed as a deadline, can be handled by the SMS alone (Response deadline > $STRT_{SMS}$). If so, no further handling is necessary. Otherwise, two more steps must be performed. In first place it is verified if the request is made over a message already present in the SMS (change of period or elimination), and, if so, a request is made to the dispatcher to eliminate the message from the synchronous message area. Next, it is evaluated if the request implies to add a message; if so, a request is made to the AMS to start its production in asynchronous mode.

The start and duration of the temporary production of synchronous messages using the AMS is controlled by the dispatcher, which sends a control message to the respective producer station to notify it about the required action. During this period of time (SSP as defined before) each station produces the required messages autonomously. The communication overhead of this control protocol is thus one message per change request.

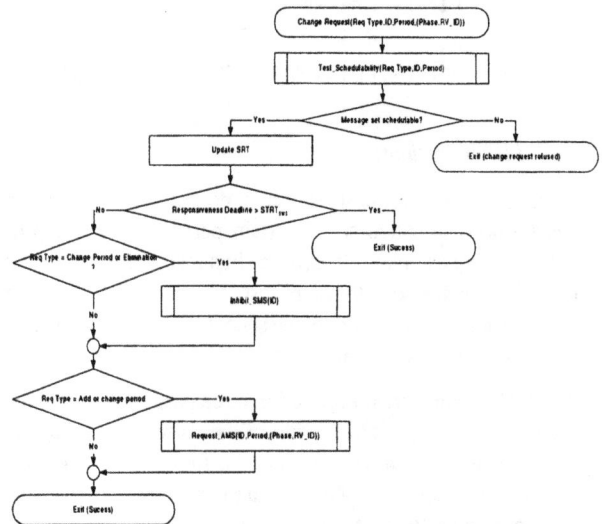

Fig. 5. Operational diagram.

The start of production message (SP_SSP) must convey the ID of the message to be produced, its period (expressed in EC's), a release delay (also in EC's) that must be applied between the reception of this message and the effective start of stream production, and the number of instances that must be produced using the AMS. Seven data bytes are used, one for variable ID, and two for message period, release delay and number of instances.

4 – ANALYSIS OF THE SYSTEM PERFORMANCE.

During the synchronous support period (SSP), the control and synchronous messages corresponding to a change request are handled by the AMS, and will compete for the bus jointly with other asynchronous messages. If timeliness guarantees are required, an evaluation should be made since the bandwidth available to the AMS is limited. For this reason, in the following subsections sufficient conditions are presented, which allow to guarantee that a set of requests is handled within specific time bounds.

4.1 Bus demand and responsiveness.

As explained in section 3, during the SSP any new and modified messages are produced using the AMS. However, if the request is accepted by the schedulability test it means that the SMS has enough leeway to hold the message. As the AMS holds the remaining bandwidth, it can be concluded that the production of data messages during the SSP will use space borrowed by the AMS from the SMS. However, this argument requires that the start of synchronous support period (SSP) takes into account the phase of the variable. This is necessary to maintain the same relative phasing in both production periods, SSP and SMS, resulting in a smooth transition from one to the other.

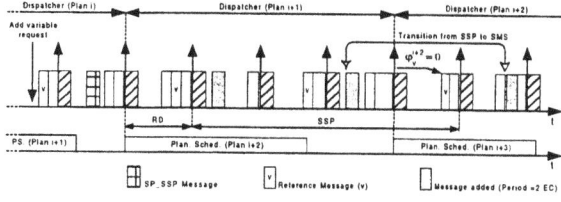

Fig. 6. Transition from SSP to SMS.

Consider for instance the example illustrated in figure 6, where a message is added with period of 2ECs and phase of 1EC relative to a reference message v. The SP_SSP message is sent by the Master Station, informing the respective producer node that it should start producing the new stream using the AMS with period of 2ECs and starting in the 2° EC after the reception of the control message. This way, the release of the first message in the stream is appropriately delayed (RD in fig. 6) so that the relative phasing is the same in SSP as in SMS.

In order to evaluate where the SSP should start, the Master node must calculate which will be the initial phase relative to the start of the plan of the first instance of the message produced in the SMS. Notice that this plan (i+2 in fig. 6) is not yet built at the request instant.

However, knowing the initial phase of a variable v on plan i, its phase in plan (i+1) is given by equation (2), where W is the length of the plan (in ECs) and P_v is the period of variable v (also in ECs).

$$\varphi_v^{i+1} = \left\lceil \frac{W - \varphi_v^i}{P_v} \right\rceil * P_v - (W - \varphi_v^i) \qquad (2)$$

When the request for a change is performed, the current scheduler instance (i+1 in fig.6) can be either terminated or still in execution. In the former case, the next plan (i+1) is already built and φ_v^{i+1} is known. Thus, equation 2 is applied once, only, to determine φ_v^{i+2}. In the latter case, plan i+1 is not built yet and thus, equation (2) must be applied twice to evaluate φ_v^{i+2} based on φ_v^i.

Knowing the relative phase of a message u with respect to a reference message v (Ph_u^v), and the initial phase of this one (φ_v^{i+2}) the number of ECs between the SP_SSP and the first instance of the message stream produced in the SMS (fig.6) is given by equation (3), where W is the length of the plan, $curEC^i$ is the EC where the request is handled within plan i ($1<=curEC<=W$) and Ph_u^v is the phase of the message being added (u) relatively to message v.

$$L_{RD+SSP_u} = W - curEC^i + W + \varphi_v^{i+2} + Ph_u^v \qquad (3)$$

Finally, the number of instances that must be produced during the SSP (NI_{SSP_u}) is given by:

$$NI_{SSP_u} = \left\lfloor \frac{L_{RD+SSP_u}}{P_u} \right\rfloor \qquad (4)$$

The release delay of the first instance relative to the reception of the control message (RD) is given by.

$$RD_{SSP_u} = L_{RD+SSP_u} - NI_{SSP_u} * P_u \qquad (5)$$

When using the AMS support to increase the responsiveness to changes in the synchronous message set, the synchronous transient response time (STRT$_{AMS}$) is substantially reduced (fig. 4). In fact, its worst-case value occurs when the request is done before the beginning of the synchronous window of one EC and the respective control message (SP_SSP) can only be transmitted in the asynchronous area of the following EC. Unless the accumulated number of control messages, due to the queuing of several requests, is greater than the available space in the asynchronous window, the STRT$_{AMS}$ will be less than 2 ECs, plus the release delay RD. Since $0 \leq RD \leq P_u-1$, the worst-case value of the responsiveness achieved by this method, expressed in ECs, is given by equation (6), where P_u is the period of variable u, measured in ECs.

$$STRT_{AMS_u} < P_u + 1 \qquad (6)$$

4.2 Pre-run-time analysis.

The SP_SSP control messages are transmitted in the asynchronous windows, competing for the bus together with other asynchronous messages. Thus, to guarantee that the bound in (6) is respected, an evaluation at pre-run-time must be performed.

As seen in the previous section, during the SSP the production of the synchronous messages is made in space borrowed from the SMS by the AMS. However, the same assumption cannot be made concerning the control messages. For these, it must be evaluated if the minimum bandwidth reserved to the AMS at configuration time (LAW – minimum length of an asynchronous window) is enough to handle them in a timely way.

In (Pedreiras *et al.*, 2000) the authors presented the analysis of the FTT-CAN Asynchronous Messaging System. Under controlled mode, due to a possible idle-time insertion (α), the effective bus time available in each EC for asynchronous transactions is less than LAW and it is given by (7).

$$LAW_{UT} = LAW - \alpha \qquad (7)$$

The inserted idle-time term (α) is intended to prevent the AMS from interfering in the SMS and is bounded by the transmission time of the longest asynchronous message (CA), which is given by equation (8), where Ca_i is the maximum transmission time of asynchronous message i and Na is the number of asynchronous messages.

$$CA = \max\{Ca_i\}, i=1..Na \qquad (8)$$

In a worst-case situation, when using either higher transmission rates or a low processing power micro-controller, the Master will take more time to handle a change request (i.e. perform the previous calculations) than to send the respective SP_SSP message. In this situation, the Master must release the bus between any consecutive SP_SSP messages. Consequently, in the meanwhile, the bus can be

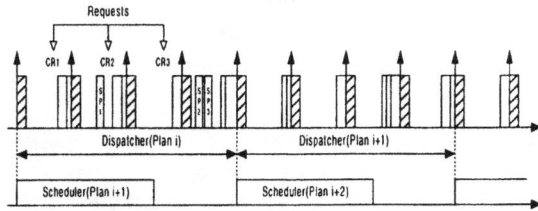

Fig. 7. Load due to SSP control messages.

taken by another asynchronous message which will cause a

blocking to the following SP_SSP message. The maximum duration of such blocking is also given by CA (8). This same blocking can happen every time the Master tries to send an SP_SSP message.

Therefore, if there are N_{CR} change requests pending, in order to guarantee that the respective SP_SSP messages can be sent in one EC so that the bound in (6) is respected, the following condition must be verified:

$$N_{CR} * [Len(SP\_SSP) + CA] <= LAW_{UT} \qquad (9)$$

This expression establishes a relationship between LAW and the maximum number of simultaneous change requests that the system is expected to handle so that the STRT of each request is still bounded by (6).

However, if there exists a minimum inter-arrival time (mit_{CR}) between the change requests, e.g. due to a serialization imposed by the schedulability analyser, the following condition results, where LEC is the length of the EC:

$$\left\lceil \frac{LEC}{mit_{CR}} \right\rceil * [Len(SP\_SSP) + CA] \le LAW_{UT} \qquad (10)$$

Finally, it is important to remind that equations (6), (9) and (10) are pessimistic. In (6) the maximum value for RD is considered. In (9) and (10) either the maximum blocking (CA) is taken into account as well as the minimum guaranteed duration of the asynchronous windows (LAW_{UT}). Therefore, the average responsiveness to change requests is greater than the values that result from using the previous equations. These are nevertheless important since they establish lower bounds to the system responsiveness.

5. CONCLUSION.

This paper discusses the levels of responsiveness demanded from communication systems in dynamic environments. In particular, it focuses on the FTT-CAN protocol, which can efficiently handle periodic (synchronous) as well as aperiodic (asynchronous) messages. However, its planning-based operation imposes some limitations to the responsiveness to requests for changes in the synchronous message set. To analyse this problem, key parameters that have impact in the responsiveness of the synchronous messaging system (SMS) of the FTT-CAN protocol are presented and their influence is discussed, namely the plan duration and the instant in which the scheduler is started.

The proposed solution to improve the SMS responsiveness,

without changing the plan length and/or the scheduler starting point, consists in using the asynchronous messaging system (AMS) to temporarily convey the changed message streams until they are taken into account by the SMS. Then, the synchronous transient response time (STRT), defined as the time lag that mediates between a change request and the instant at which the corresponding new requirements are reflected in the bus traffic, is substantially reduced. Such time lag is upper bounded by the period of the variable plus one EC. If the schedulability test is performed, its execution time adds to this bound.

Finally, two sufficient conditions are established which allow to define at pre-runtime the required minimum bandwidth for the AMS so that a given number of change requests, either simultaneous or separated by a minimum inter-arrival time, can be timely handled.

REFERENCES

Almeida, L (1999). Flexibility and Timeliness in Fieldbus-based Real-Time Systems. PhD Thesis. University of Aveiro-Portugal,

Almeida, L., J.A. Fonseca, P. Fonseca (1999). A Flexible Time-Triggered Communication System Based on the Controller Area Network: Experimental Results. *Proc. of FeT'99 (Int. Conf. on Fieldbus Technology)*. Magdeburg, Germany.

Fohler, G. (1995). Joint Scheduling of Distributed Complex Periodic and Hard Aperiodic Tasks in Statically Scheduled Systems. *Proc. 16th Real-Time Systems Symposium.* Pisa, Italy.

Kopetz, H. (1997). *Real-Time Systems Design Principles for Distributed Embedded Applications.* Kluwer Academis Publishers.

Pedreiras, P. , L. Almeida (2000). Combining Event-triggered and Time-triggered traffic inFTT-CAN: Analysis of the Asynchronous Messaging System. *Proc. WFCS 2000,* Oporto.

Peraldi, M.A. and J.D. Decotignie (1995). Combining Real-Time Features of Local Area Networks FIP and CAN. *Proc. of ICC'95 (2nd Int. CAN Conference), CiA – CAN in Automation.*

Copyright © IFAC Distributed Computer Control Systems,
Sydney, Australia, 2000

A DISTRIBUTED FAULT-TOLERANT REAL-TIME SKELETON

Matthew E. Outram[*,1] Fethi A. Rabhi[**]

*Department of Computer Science, The University of Hull, Hull
HU6 7RX (UK)*
**School of Information Systems, The University of New South
Wales, Sydney 2052 (Australia)*

Abstract: This paper is concerned with high-level design techniques for distributed fault-tolerant real-time systems. Based on the concept of skeletons, a design environment which supports fault-tolerance as a set of predefined interaction patterns is proposed. The environment supports three main features. Firstly a graphical user interface is available for the specification of predefined problem specific parameters. Secondly the specified parameters may be analysed, both fault-tolerance and cost measures extracted and a static distributed schedule determined. Finally the proposed environment can automatically generate distributed code in C and PVM for testing purposes. *Copyright ©2000 IFAC*

Keywords: Software development, automatic analysis, skeletons, fault-tolerance

1. INTRODUCTION

The design of a distributed fault-tolerant system presents a challenge to the software designer, since all the following issues must be considered at once:

- correctness: in the presence of concurrent processes, dealing with process synchronisation and the prevention of live-lock and dead-lock.
- scheduling: determining the right order of executing processes within one processor such that timing and resource constraints are satisfied.
- fault-tolerance: choices have to be made according to the strategy used. For example, N-version programming requires the selection of a suitable voting strategy (i.e. determining the correct output from a choice of N alternatives).

- distribution: allocation of processes to processors while satisfying timing, performance, reliability and resources constraints.

While the first three issues have been extensively studied (Burns and Wellings 1997), the last one introduces new problems as distribution decisions often cannot be made without making various compromises with earlier choices. For instance, it is important for a fault-tolerant system to adhere to a set of rules concerning load balancing. For instance (and as will be shown by the metric later), it is desirable for certain processes to be clustered together whereas it is advantageous for other processes to be scheduled onto different processors. Consider the following example: assume two processors (N_1 and N_2) and four processes ($P_1 \ldots P_4$). Further assume that P_2 is a replication of P_1 and likewise P_4 a replication of P_3 and that at least one of each replication is needed for system success. For numerical problems with even computation and communication a suitable load-balance is as shown in figure 1. This arrangement is suitable in terms of computation and commu-

---

[1] Current address: Fujifilm Electronic Imaging Ltd, Boundary Way, Hemel Hemstead HP2 7RH (UK)

nication (i.e. evenly loaded). In terms of fault-tolerance, figure 1 has a low threshold as a single failure in either processor results in overall system failure. Fault-tolerance requires the *role* of the process to be considered alongside the computation and communication of each process. Figure 2 shows a similar load-balance but this time with a much better degree of fault-tolerance (either of the processors may fail and the system continues to operate).

This paper describes an approach which could address all these issues at a much earlier stage in the development process and proposes a programming environment based on skeletons, an idea borrowed from parallel programming where a system is built using predefined interaction patterns. Its main contribution is in providing a general overview of the proposed environment, describing its graphical user interface and explaining how key design decisions are supported. The paper also briefly mentions the metrics used and the code generation process.

2. A SKELETON FOR DISTRIBUTED FAULT-TOLERANCE

2.1 *What is a Skeleton?*

The philosophy of the *skeleton* approach (Cole 1989) is to abstract out from the development process some of the aspects of a parallel algorithm (i.e. communication, synchronisation, load-balancing etc.). Since this concept has been used in various forms and contexts, we are primarily interested in the idea of using a skeleton as the basis of a parallel programming environment (Rabhi 1995). In such an environment, the basic structure of the skeleton, for which efficient algorithms already exist, is predefined. The user then expresses the problem specific parameters and in doing so "customizes" the skeleton to the particular application requirements. This allows for environments that combine high-level interfaces for parameter specification, skeleton properties analysis, and automatic code generation. Examples of such environments can be found in (Osoba and Rabhi 1998) and (Parsons and Rabhi 1998).

2.2 *Proposed Approach*

The proposed Multiple Processor Fault-Tolerant Skeleton (MPFTS) considers applications which fit the feedback model and require fault-tolerance through the replication of processes and their distribution across multiple processors. Figure 3 shows an overview of the experimental system.

A skeleton-based approach has been chosen for many reasons. One such reason is that of intellectual abstraction, since by their very nature skeletons hide the parallel (or distributed) decomposition and distribution (and hence communication and synchronisation) of an algorithm. The user interface is described next and it is here that the user is provided with the opportunity to "customize" the skeleton with the problem specific parameters. Furthermore the user interface provides access to the scheduling analyser which is detailed in the next section. Section 5 describes the automatic code generation provided by the *MPFTS* Builder which generates the target code for a range of specific architectures.

3. THE USER INTERFACE

The User Interface offers the user a graphical method for the definition of the parameters of the skeleton. Figure 4 shows a decomposition of the "Problem Specific Parameters" shown in figure 3. The remainder of this section describes these parameters and, where appropriate, gives details of the experimental system.

At the top level the user is able to select a fault-tolerance strategy suitable for the current problem. In this work, we only consider fault-tolerance achieved by the use of redundancy. In software, this is called *N-version* programming. This technique simply maintains several copies of the controlling system (called the *controller* process) and all outputs are subject to a voting mechanism (by a *voter* process) to determine eventual errors. To improve reliability, voter processes can themselves be duplicated. A comprehensive overview of fault-tolerance can be found in (Pradhan 1996).

3.1 *Cost Modelling and Scheduling*

These parameters provide the system builder with a view of the implications that the selection of an architecture has upon the scheduling of a real-time system.

There are window interfaces which allow:

- **architecture selection**: choosing the characteristics of the target architecture i.e. type, number of nodes, topology (see figure 5). At the moment, all architectures must support the PVM library.
- **cost parameters** : the user can define either a *cost level* or a *fault-tolerance level* using "knobs" (scale widgets) as shown in figure 5. Once one level is set, setting the other one automatically is achieved by pressing the *calculate* buttons as appropriate. Details

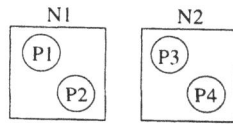

Fig. 1. First Example of a Schedule

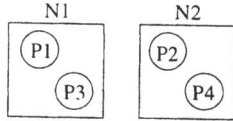

Fig. 2. Second Example of a Schedule

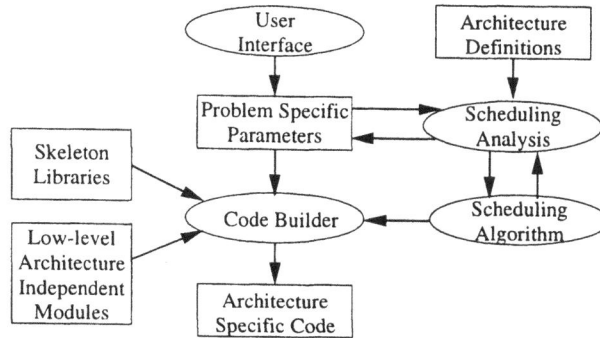

Fig. 3. Overview of the Experimental System

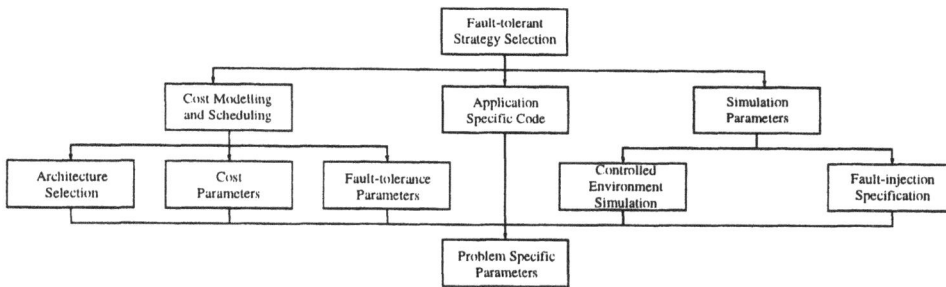

Fig. 4. Overview of the Problem Specific Parameters

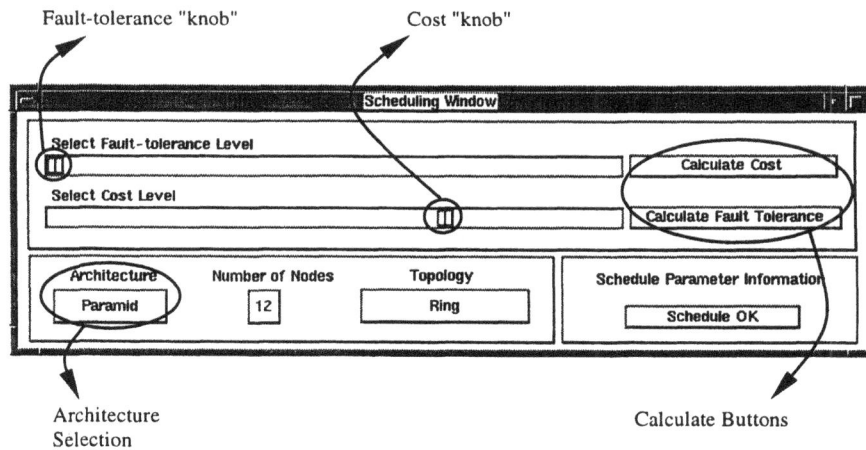

Fig. 5. Interface: Cost Modelling and Scheduling

about these parameters will be provided in Section 4.

- **fault-tolerance parameters**: since N-version programming is used, the user needs to specify the number of replicated controllers and the number of replicated voters. Figure 6 shows the replication management facility configured for a case study with 3 controllers and 3 voters.

3.2 *Application Specific Code*

Figure 7 shows the window which allows for the specification of the *application specific code*. This window uses two reserved arrays; inputs and outputs which respectively represent the inputs into the controlling system and the outputs to the controlled environment. The relationship between the inputs and the outputs in the controlling system is expressed procedurally as illustrated on the top left of figure 7. The experimental system uses a C-like syntax which gives the advantage of familiarity and portability. Some C programming constructs are prohibited as they make the program nondeterministic (which makes temporal analysis impossible). Such constructs include dynamic data structures, recursion and unbound loop structures. Initial input and output values are specified in the right hand-side of the window.

It is important to note that only the control part of the software needs to be defined, the programmer does not need to know anything about the replication of controllers and the voting mechanism.

3.3 *Simulation Parameters*

To enable testing of the generated code, it is possible to emulate the controlled environment by defining relationships between the outputs and the inputs. This is shown at the bottom left of figure 7. There is also a fault-injector facility which enables the system to be tested in the presence of faults. The experimental system restricts these faults to timing faults in either the voting sub-system or the controller sub-system. The output from a simulation is displayed by a display process which acts as the output device associated with an instrumentation panel. In the experimental system output information is in a text based format.

4. SCHEDULING METRICS

We now define the metrics used by the system to decide the best allocation of processes to processors. As the example in the introduction illustrated, there are two requirements to satisfy:

- Minimising cost: this corresponds to obtaining the most efficient schedule
- Maximising fault-tolerance: this corresponds to obtaining the most reliable schedule

We have seen that the user can specify one or the other by setting the corresponding knob to a maximum through the user interface (see Section 3.1). The cost level knob corresponds to the first requirement (measure C) and the fault-tolerance level knob corresponds to the second requirement (measure R). Alternatively, the user can set the level to an intermediate value if a compromise is sought. The scheduling algorithm used is based on a list heuristic (Price and Salama 1990). The heuristic uses a priority assignment scheme based on communication volume with the aim of reducing overall communication costs. The general list heuristic algorithm is extended to incorporate the idea of Fault-Tolerant clustering. More details about the metrics used and the scheduling algorithm are provided in (Outram 1999).

5. CODE GENERATION

The system provides a compiler which reads a parameters file containing all the information supplied by the user and generates target code using the Skeleton Libraries and the Low-level Architecture Independent Modules (see figure 3). In the experimental system, the system modules are coded in PVM and the target code is C code which can be compiled with the "make" facility. A *makefile* for each architecture is available for this purpose. Further details are provided in (Outram 1999).

6. RELATED WORK

Related work can be classified in two distinct categories: high-level design environments for real-time systems and those for parallel systems. Examples of the former include StateMate (Harel *et al.* 1990) and HRT-HOOD (Burns and Wellings 1995). HRT-HOOD has a comprehensive repertoire of design constructs which allow both cyclic and asynchronous processes and a wide range of communication protocols to be modelled. In addition, the design activity is enforced by a range of rules to enable dependability analysis. Compared to our system, these approaches suffer from the fact that they cannot deal with large numbers of processes arranged in replicated structures. For example a user has to explicitly specify all aspects related to fault-tolerance and changes in the number of replicated modules, for instance, imply a complete redesign of the system. Methodologies that provide support for real-time systems as well as replicated structures include PARSE (Gorton

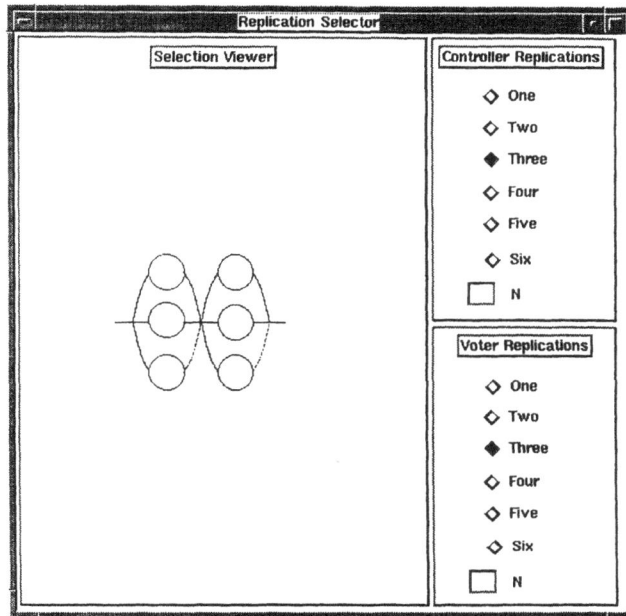

Fig. 6. Interface: Replication Management

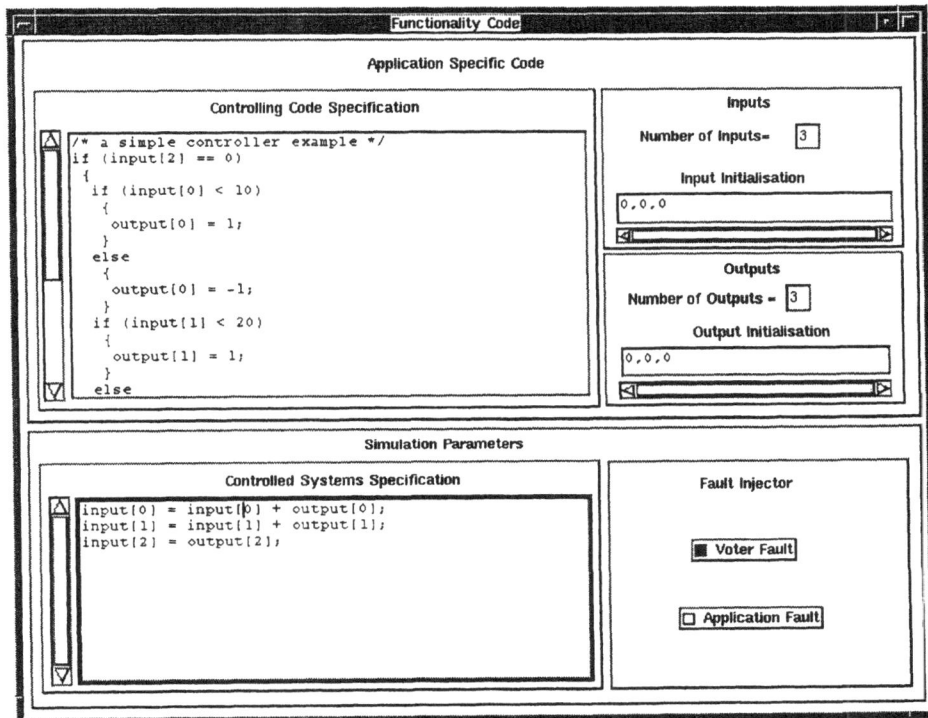

Fig. 7. Interface for Application-Specific Information

105

et al. 1995) and ADL (Polman and van Steen 1996).

High-level parallel programming environments (particularly those skeleton-based) are much better in providing abstractions for process replication and regular communication structures. However, their scope has been much restricted to data-parallel structures such as in (Osoba and Rabhi 1998, Parsons and Rabhi 1998).

A closely related project is the *ReSoFT* system described in (Tso *et al.* 1997). This system provides a mechanism to build systems from (user-constructed) libraries of reusable software components. The system provides the choice of two Fault-Tolerant techniques; N-version programming and Recovery Blocks. However, the system is very much tied to a specific language (Ada95) and a particular architecture which is a network of workstations connected via a redundant Ethernet network.

7. CONCLUSION AND FUTURE WORK

The paper proposed a Multiple Processor Fault-Tolerant Skeleton (MPFTS) which is specifically designed for applications which fit the feedback model (section 1) and which require fault-tolerance through the replication of processes and their distribution across multiple processors. The rationale behind such a proposed environment is to bring forward fault-tolerance aspects during the development process, provide cost measures for scheduling, and finally support a mechanism for the automatic production of code based on techniques that have already had empirical success.

One area of future work is to develop a base-language more suitable for the real-time systems domain. Future work is also needed to consider how the system could be adapted to make optimal predictions based upon partial application specification. For example, it would be better if the experimental system was able to suggest an optimal architecture or an optimal number of replications to the user. Another area would be to extend the skeleton to include a new class of parameters; *timing parameters*. These parameters could be specified to give more specific deadlines to the system. The skeleton could also be extended to incorporate other models of fault-tolerance, such as roll-back or roll-forward techniques. The mechanisms for interfacing these models with the application specifics should provide an interesting challenge.

8. REFERENCES

Burns, A. and A.J. Wellings (1995). *A Structured Design Method for Hard Real-Time Ada Systems*. Elsevier.

Burns, A. and A.J. Wellings (1997). *Real-Time Systems and Programming Languages (2nd Ed.)*. Addison-Wesley.

Cole, M. (1989). *Algorithmic skeletons: a structured approach to the management of parallel computation*. Research monographs in Parallel and Distributed Computing, Pitman.

Gorton, I., J. Gray and I. Jelly (1995). Object-based modelling of parallel programs. *IEEE Parallel and Distributed Technology* **3**(2), 52–63.

Harel, D., H. Lachover, A. Naamad, A. Pnueli, M. Politi, M. Sherman, A. Shtull-Trauring and M. Trakhtenbrot (1990). STATEMATE: a working environment for the development of complex reactive systems. *IEEE Transactions on Software Engineering* **16**(4), 403–414.

Osoba, B.O. and F.A. Rabhi (1998). A parallel multigrid skeleton using bsp. In: *Proceedings of EuroPar'98: Parallel Processing, Lecture Notes in Computer Science 1470* (D. Pritchard and J. Reeve, Eds.). Springer Verlag. pp. 704–708.

Outram, M. (1999). A skeleton-based programming environment for parallel and distributed fault-tolerant real-time systems. Technical Report PhD Thesis. Department of Computer Science, The University of Hull. Hull.

Parsons, P.J. and F.A. Rabhi (1998). Generating parallel programs from paradigm-based specifications. *Journal of Systems Architecture* **45**(4), 261–283.

Polman, M. and M. van Steen (1996). A structured design technique for distributed programs. Technical Report EUR-FEW-CS-96-04. Department of Computer Science, Erasmus University. Rotterdam.

Pradhan, D.K. (1996). *Fault-tolerant computer systems design*. Prentice Hall.

Price, C.C. and M.A. Salama (1990). Scheduling of precedence-constrained tasks on multiprocessors. *The Computer Journal* **33**(3), 219–229.

Rabhi, F.A. (1995). Parallel programming methodology based on paradigms. In: *Transputer and Occam Developments* (P.A. Nixon, Ed.). pp. 239–252. IOS Press.

Tso, K.S., E.H. Shokri and R.G. Dziegiel (1997). Resoft: a reusable testbed for development and evaluation of software fault-tolerant systems. In: *High Assurance Systems Engineering Symposium*. Washington D.C.

Copyright © IFAC Distributed Computer Control Systems,
Sydney, Australia, 2000

RTP/RTCP BASED REAL-TIME PROTOCOL OVER ETHERNET FOR DISTRIBUTED CONTROL SYSTEM

Joonwoo Park ** Jaehyun Park ***

*Automation Engineering, Inha University,
253 Younghyundong, Inchon, 402-751 Korea.*
*** e-mail: jwoo@rcsl.inha.ac.kr*
**** e-mail: jhyun@inha.ac.kr*

Abstract: This paper proposes an extended RTP/RTCP (real-time transport protocol/RTP control protocol) that can be applied to the distributed control systems connected by Ethernet. E-RTP/RTCP is designed to be independent of the underlying transport and network layers, and supply the soft real-time communication to the control systems without operating system kernel modifications. E-RTP/RTCP also provides PTM/RTP (Periodic Transmission Mode over RTP) for the periodic data transmission that is useful for the digital control systems. This paper includes the computer simulation and implementation results of E-RTP/RTCP. Copyright© 2000 IFAC

Keywords: Real-time Ethernet, RTP/RTCP, Distributed Control System(DCS), Quality-of-Service (QoS)

1. INTRODUCTION

A communication network is one of the most important parts to provide the real-time performance for the whole distributed computer control systems (DCCS). In the past, expensive field-buses were used to guarantee the real-time data exchange between the control devices. However, distributed control systems using these kinds of field-buses have several shortcomings: (1) expensive to implement, (2) hard to code application softwares, and (3) there is no world wide standard even there are many application-specific standards, including Profi-bus, World FIP, Foundation Fieldbus, and CAN.

Meanwhile, as the Internet technology progresses, the network protocols for Internet, TCP/IP and Ethernet, are likely to be used for the control area too. TCP/IP and Ethernet have many strong points for the control application; inexpensive to implement, vendor neutral worldwide standards, and compatible with large selection of application softwares.

Even TCP/IP and Ethernet protocols are attractive to the control system designers, they are not yet easily used for the real-time distributed control systems directly because the real-time communication between the control nodes is not guaranteed. This non-real-time characteristics is mainly caused by packet collisions on the Ethernet channel and unnecessary long delay in TCP/IP protocol layer. Figure 1 shows a typical distributed control system that consists of a controller, sensor and actuator nodes, and they are connected by a computer network. The indeterministic transmission delay between the nodes

* The work reported in this paper was supported in part by Development of Semiconductor Equipment Communications Standards Project (G7 Project Grant No: 5-3-1-4). Any opinions, findings, recommendations, and conclusions expressed in this paper are those of the authors and do not necessarily reflect the view of the funding agencies.

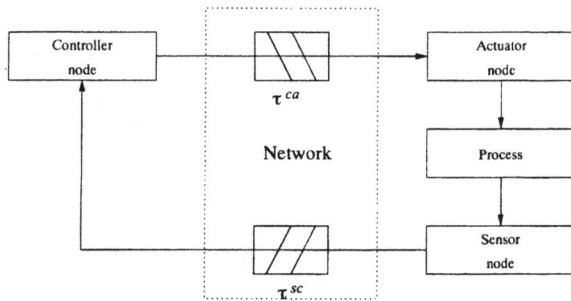

Fig. 1. Simple DCS with network delays.

contributes as one of the major disturbance to the control loops, and in turn, it affects stability of the control system(Nilsson *et al.*, 1998; Kweon *et al.*, 1999).

The efforts to use TCP/IP and Ethernet protocols for real-time distributed control systems are categorized into several approaches. First approach is to modify MAC(Medium Access Control) layer of Ethernet to avoid the packet collision problem that causes the unpredictable delay(Tanenbaum, 1989). BRAM (Broadcast Recognizing Access Method)(Chlamtac *et al.*, 1979) and MBRAM (Modified BRAM) are the protocols that resolve the problem(Signorile, 1988). This approach, however, has some problems to be widely used: (1) no standard contention-free MAC protocol is fixed yet and (2)all nodes connected to physical media should support the same scheduling algorithm. Even single node that uses standard MAC protocol may cause unpredictable congestion problems. These reasons make it difficult to use modified MAC protocol for the real-time systems.

Second approach is to modify the mechanism of TCP layer. TCP/IP was originally designed to be used in the inter-networking environments where long latency and relatively unreliable data without guaranteeing deadline are expected(Stevens, 1994). Therefore it has a lot of unnecessary overhead for a local area network (LAN). A series of algorithms and implementation techniques to improve the performance of the transmission control protocol (TCP) have been proposed: TCP for the transactions(Braden, 1992; Braden, 1994), receiving overhead reduction technique(Clark *et al.*, 1989), header prediction algorithm(Jacobson, 1990), and LAN/TCP for Ethernet only(Park and Yoon, 1998). But these protocols need modifying of the operating system kernel, and are hardly to provide compatibility.

Third approach is to use RTP (Real-time Protocol) that is originally invented for the real-time communication of multimedia packet (Audio-Video Transport Working Group, 1996). RTP provides unidirectional periodic end-to-end delivery services. Those services include a payload type identification, sequence numbering, time stamp-

ing, and delivery monitoring. RTP and multimedia applications are generally run on top of connectionless protocol because somehow packet loss is allowable in the multimedia applications.

Recent approach is a research on the flow labeling protocol of IP version 6 (IPv6). The flow label field in a IPv6 header may be used for supporting a real-time feature and guaranteeing QoS(Partridge, 1995; Deering and Hinden, 1998). The feature, however, is useful in WAN (Wide Area Network) environment with multihops rather than in LAN (Local Area Network) environment where a usual DCCS is.

This paper proposes an extended RTP/RTCP (real-time transport protocol/ RTP control protocol), E-RTP/RTCP, which can be applied to the distributed control system over Ethernet.

The outline of the rest of this paper is as follows. Section 2 gives a short introduction to RTP/RTCP and their problems in applying to DCS. Section 3 describes the proposed E-RTP/RTCP and their key features. Section 4 presents the simulation results for E-RTP/RTCP.

2. RTP/RTCP

RTP/RTCP protocols are originally invented for the multimedia applications such as digitized voice and video transmission using multicast or unicast network services. Since the control applications, however, differ from the multimedia applications in handling the real-time data, RTP/RTCP itself is not adequate for the control applications in the distributed control systems. To clarify the issues in applying RTP/RTCP to the DCS applications, the characteristics of RTP and RTCP are introduced in this section.

2.1 *RTP(Real-time Transport Protocol)*

RTP is proposed to provide end-to-end transport services for the real-time multimedia data in multicast or unicast manner. Even RTP itself does not provide any mechanism to ensure timely delivery of data or any guaranteed quality-of-service, it supports a rate-based flow control, that controls data flow explicitly. This rate-based flow control is a very efficient method for the multimedia data that should be transmitted at a constant rate. In case of packet loss, however, it is very hard to retransmit the lost packet because the rate feedback data of RTP totally depends on the RTCP message that is periodic also. Although some packet loss is allowed in the multimedia applications, it may usually cause a serious problem in the industrial control applications. As a result, RTP itself needs some kind of modifications not

Table 1. Additional payload types defined in E-RTP/RTCP.

payload type	definition name	PTM/RTP	early start	NACK
80	E-RTP(normal)	x	x	x
81	E-RTP(request early start)	x	o	x
82	E-RTP(early start on PTM)	o	o	x
83	E-RTP(with NACK)	o	x	o
84	E-RTP(with early start and NACK)	o	o	0

to miss control packets for the real-time control applications.

2.2 *RTCP(RTP control protocol)*

Real-time Transport Control Protocol(RTCP), usually used with RTP, is a control messages for the RTP flow control. It transmits the periodic control packets to all participants in the session, using the same distribution mechanism as the RTP does. RTCP packet contains the network status including packet delay, jitter, packet counts, that are important information for the rate-based flow control algorithm of RTP. Though the periodically transmitted data of RTCP is inevitable for the dynamically changing multimedia environments, it is a big overhead in the control environments where network configuration does not change often. Furthermore, the periodic TCP data could not reflect network status immediately, retransmission of the lost packet is not always possible within the deadline. This means RTCP also has problems to be used for the real-time control applications directly.

3. EXTENDED RTP/RTCP

This section extends the original RTP/RTCP for the real-time control applications in the DCS environments. The basic idea of the *Extended RTP/RTCP*(E-RTP/RTCP hereafter) is (1) to change the transmission method of RTCP from periodic-base to event triggered-base, and (2) to support a periodic data communication, *Periodic Transmission Mode for RTP*, (PTM/RTP hereafter), with less network overhead than the normal RTP/RTCP. To maintain the backward compatibility with normal RTP/RTCP, E-RTP/RTCP header follows the normal RTP/RTCP header structure but new payload types are introduced as defined in Table 1.

3.1 *Event triggered RTCP*

Event triggered RTCP proposed in this paper is triggered not periodic-base but events-base when one of the following conditions happens:

- Whole system is initialized.

Table 2. A summary of notations

symbol	Description
τ^{ca}	Communication delay between the controller and the actuator.
τ^{sc}	Communication delay between the sensor and the controller.
τ_n	Response time of normal RTP
τ_p	Response time of PTM/RTP
D_{sys}	System process delay.
T_{sys}	System sampling period.
T_{rtp}	RTP timer period.

- Members participate in/depart from a session.
- Nodes require to increase/decrease the transmission period.
- Nodes detect beyond permissible packet loss or network latency.

About 5% of RTP bandwidth is comsumed for the periodic RTCP messages when normal RTP/RTCP protocol is used. The proposed event-based RTCP, however, rarely occupies network bandwidth, unless all members dynamically participate in/depart from a session. The proposed protocol can reflect the network status immediately, when detecting beyond permissible packet loss or network latency.

3.2 *Periodic Transmission*

Periodic data exchange between the control nodes are fundamental data transfer activities in DCS system. E-RTP/RTCP includes a special scheme for reliable periodic transfer based on RTP called PTM/RTP. To reduce transmission latency and to improve network utilization, *Quick Response* and *NACK-based Retransmission* are used.

Quick Response Normal RTP packet are transmitted periodically at the pre-defined time interval, that can be adjusted by RTCP control message. This periodic data transmission is adequate for the multimedia application because in those applications, data usually flows unidirectional(downstream or upstream only). In control applications, however, data flows bidirectional, usually control reference and sensory data pair. The periodic RTP packet imposes an excessive latency between this data pair. To reduce a network latency for the control applications, the PTM/RTP provides *Quick Response* scheme with which RTP packet is responded

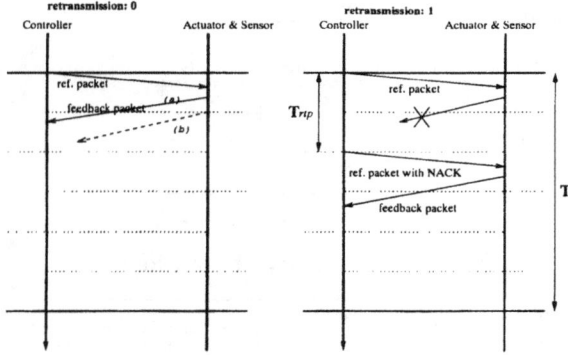

Fig. 2. time line for PTM/RTP transmission and retransmission.

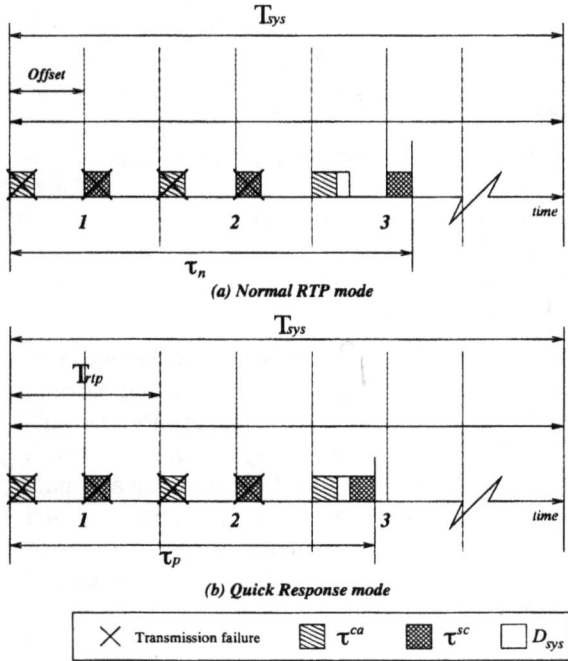

Fig. 3. PTM/RTP transmission scheme.

immediately after RTP packet received without waiting for next RTP interval. This Quick Response is especially useful for the retransmission of the lost or damaged packet. Figure 2 shows the time line of the PTM/RTP with Quick Response.

Figure 3 compares normal RTP with the PTM/RTP. In this figure, T_{rtp} should be selected as

$$\tau_k^{ca} + D_{sys} \leq \frac{T_{rtp}}{2}. \qquad (1)$$

And response time t_n and t_p can be calculated as

$$\tau_n = \frac{T_{rtp}}{2} + (k-1) * T_{rtp} + \tau_k^{sc}, \qquad (2)$$

$$\tau_p = (k-1) * T_{rtp} + \tau_k^{ca} + D_{sys} + \tau_k^{sc}. \qquad (3)$$

Retransmission Algorithm Because there is no retransmission algorithm in RTP, the loss or damaged packet cannot be recovered. Because in the control application, missing single

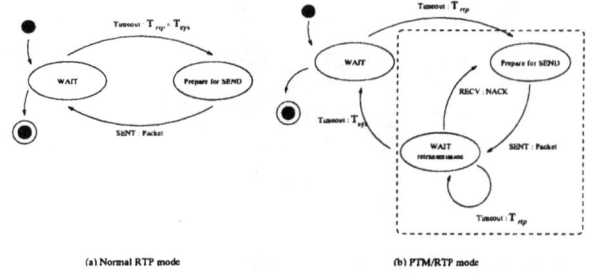

Fig. 4. PTM/RTP state transition diagram.

control packet may cause a serious problem while it dose not in multimedia applications, a retransmission algorithm should be provided to use RTP for the control applications. This paper uses two RTP channels for bidirectional data exchange. When a packet is lost, *negative acknowledge(NAK)* is notified. However, the retransmission to the NAK packet may misses the deadline because all the packet exchange happens at the RTP time interval. To overcome this deadline-miss problem, an oversampling method is used. RTP time interval, T_{rtp}, can be set to the internal several times faster than control system's sampling time. This time-redundancy helps meet the deadline even a lost or damaged packet happens.

One drawback of the oversampling is that it increases the network load and, as a result, the packet collisions in the Ethernet. This may be a serious problem when a large number of nodes participate in the session. To prevent network congestion caused by excessive data packet, proposed algorithm uses a NACK-based retransmission scheme with the restriction of retransmission counts.

Figure 4 shows the state machine of E-RTP/RTCP's retransmission scheme, where additional timer, T_{rtp}, controls retransmission state to meet the deadline.

4. PERFORMANCE EVALUATION

The real-time performance of the proposed protocol is evaluated by computer simulation in this section. Arena, a widely used simulation package(Kelton *et al.*, 1997), is used for modeling the whole DCS environments connected by Ethernet. The simulation model includes the CSMA/CD MAC protocol of Ethernet, TCP/IP protocol stacks, and the proposed E-RTP/RTCP to check the overhead and performance of the protocol stacks. For the precise evaluation of the indeterministic packet delay in the Ethernet layer, the binary back-off algorithm, recommended by IEEE 802.3, rather than random delay is used for the packet collision model.

The performance is evaluated in terms of the transmission latency, the number of deadline-

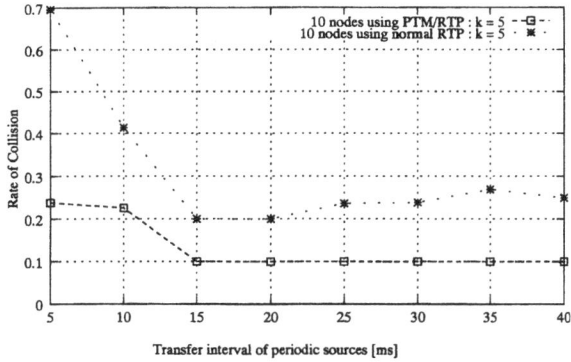

Fig. 5. Rate of collision versus transfer interval.

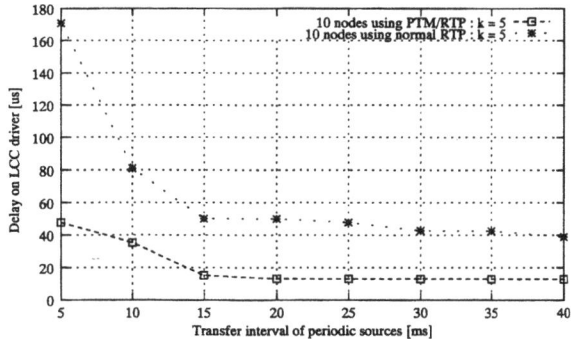

Fig. 6. Transmission delay versus transfer interval.

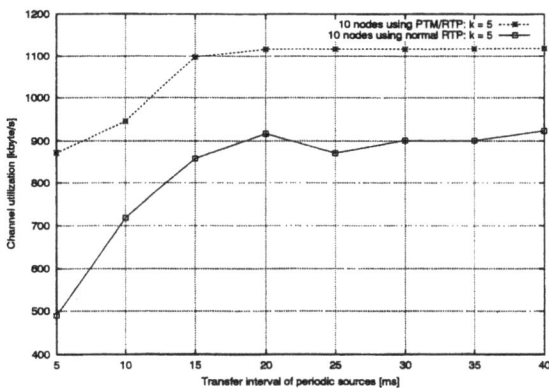

Fig. 7. Channel utilization versus transfer interval.

miss-packets, and the network utilization with varying the number of DCS nodes and transmission intervals between the control and sensory packets. In the simulation model, time line is slotted into a bit-time unit, 100 nano second at 10 MBPS. Transmission can start only at the beginning of a slot, which is a feasible assumption because slot is narrow enough. To get the average latency, 100,000 messages of 128 bytes(1024 bits) long were generated and the latency of them were measured.

Figure 5, 6 and 7 show collision rate, average transmission delay, and channel utilization respectively, for the standard RTP and E-RTP with same network configuration. The collision in the Figure 5 is the number of collisions per transmission, that saturates at 0.1 when transmission

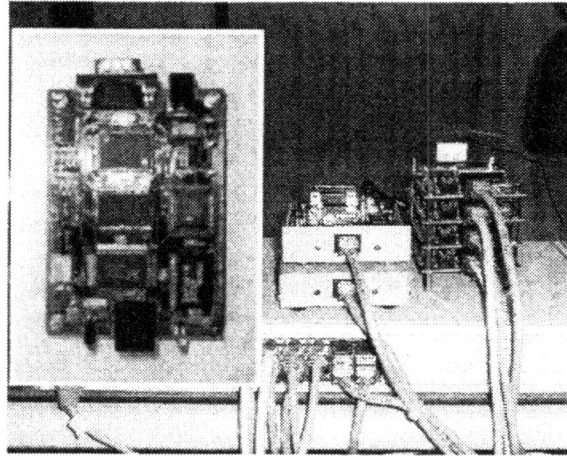

Fig. 8. Photograph of the experiment system.

interval is longer than $15ms$. Figure 6 shows that the transmission latency caused by LCC driver is also astringent below $20\mu s$. Figure 7 shows the PTM/RTP is about 22% higher in channel utilization, compared to that of the RTP/RTCP.

5. IMPLEMENTATION

Since the proposed protocol is an application layer protocol, it provides soft real-time communication without the modification of operating system kernel. For implementation and experiments environments, an experimental system is implemented using AM188ES-based single board controllers with a simple real-time O/S that has BSD-style TCP/IP kernel and E-RTP/RTCP shown in Figure 8. For monitoring and implementing the host controllers, E-RTP/RTCP is also implemented onto the Linux and Solaris boxes. The experimental system consists of six single board controllers and two Linux boxes.

6. CONCLUSION

This paper proposes an application layer protocol, Extended RTP/RTCP, for the soft real-time communications in the distributed computer control systems environments. The proposed protocol, E-RTP/RTCP, is based on the RTP/RTCP that is originally designed for the multimedia real-time data, and provides stochastic real-time communication for the control applications. It also supports a special periodic data transmission feature, PTM/RTP, with less overhead than the original RTP/RTCP, that is frequently used for the control and sensor data data transmission in the distributed control systems. The computer simulation and experiment results show that the proposed scheme economically improves the communication latency and network utilization, compare to the existing approaches.

7. REFERENCES

Audio-Video Transport Working Group (1996). RTP: A transport protocol for real-time applications. *RFC1889*.

Braden, R. (1992). Extending TCP for transactions – concepts. *RFC1379*.

Braden, R. (1994). T/TCP – TCP extensions for transactions functional specification. *RFC1644*.

Chlamtac, Imrich, William R. Franta and K. Dan Levin (1979). BRAM: The broadcast recognizing access method. *IEEE Transactions on Communications* **27**(8), 1183–1189.

Clark, David D., Van Jacobson, John Romkey and Howard Salwen (1989). An analysis of tcp processing overhead. *IEEE Communication Magazine* pp. 23–29.

Deering, S. and R. Hinden (1998). Internet protocol, version 6(IPv6). *RFC2460*.

Jacobson, V. (1990). 4BSD TCP header prediction. *Computer Communication Review*.

Kelton, W. David, Randall P. Sadowski and Deborah A. Sadowski (1997). *Simulation with ARENA*. McGraw-Hill.

Kweon, Seok-Kyu, Kang G. Shin and Qin Zheng (1999). Statistical real-tiem communication over ethernet for manufacturing automation systems. *Proceedings of the 5th Real-Time Technology and Applications* pp. 192–202.

Nilsson, J., B. Bernhardsson and B. Wittenmark (1998). Stochastic analysis and control of real-time systems with random time delays. *Automatica* **34**(1), 57–64.

Park, Jaehyun and Youngchan Yoon (1998). An extended TCP/IP protocol for real-time local area networks. *IFAC Control Engineering Practice* **6**(1), 111–118.

Partridge, C. (1995). Using the flow label field in ipv6. *RFC1809*.

Signorile, Robert P. (1988). MBRAM-A priority protocol for PC based local area networks. *IEEE Network* **2**(4), 55–59.

Stevens, W. Richard (1994). *TCP/IP Illustrated, Volume 1*. Addison-Wesley Publishing Company. Massachusetts.

Tanenbaum, Andrew S. (1989). *Computer Networks*. Prentice Hall, Inc.. Englewood Cliffs, New Jersey.

Copyright © IFAC Distributed Computer Control Systems,
Sydney, Australia, 2000

DISTRIBUTED COMPUTER-CONTROLLED SYSTEMS: THE DEAR-COTS APPROACH [1]

P. Veríssimo*, A. Casimiro*, L. M. Pinho**, F. Vasques***, L. Rodrigues*, E. Tovar**

*University of Lisboa, FCUL, Bloco C5, Campo Grande,
1749-016 Lisboa, Portugal, E-mail: pjv,casim,ler@di.fc.ul.pt
** Polytechnic Institute of Porto, ISEP, Rua de São Tomé,
4200-072 Porto, Portugal, E-mail: lpinho,emt@dei.isep.ipp.pt
*** University of Porto, FEUP, Rua Dr. Roberto Frias, 4200-465
Porto, Portugal, E-mail: vasques@fe.up.pt

Abstract: This paper proposes a new architecture targeting real-time and reliable Distributed Computer-Controlled Systems (DCCS). This architecture provides a structured approach for the integration of soft and/or hard real-time applications with Commercial Off-The-Shelf (COTS) components. The Timely Computing Base model is used as the reference model to deal with the heterogeneity of system components with respect to guaranteeing the timeliness of applications. The reliability and availability requirements of hard real-time applications are guaranteed by a software-based fault-tolerance approach. *Copyright © 2000 IFAC*

Keywords: Real-Time, Fault Tolerance, Distributed Embedded Systems, COTS

1. INTRODUCTION

Currently, there is a trend to incorporate Commercial Off-The-Shelf (COTS) components in Distributed Computer-Controlled Systems (DCCS), in order to minimise costs and development time. Therefore, there is the need for architectures allowing the integration of COTS components, hard real-time reliable applications and non-reliable and/or soft real-time applications.

This paper proposes the DEAR-COTS (Distributed Embedded ARchitecture using Commercial Off-The-Shelf components) architecture for supporting real-time and reliable DCCS. This architecture provides DCCS with a generic framework, in which hard real-time applications can execute, guaranteeing their timeliness and reliability requirements, whilst, at the same time, embodying soft real-time or non real-time applications.

This framework is put together by means of a Timely Computing Base (TCB), which is used as the reference model to deal with the heterogeneity of system components with respect to guaranteeing the timeliness of applications.

The remainder of this paper is organised as follows. Section 2 presents some of the relevant work concerning dependable distributed real-time systems. Afterwards, Section 3 presents the proposed architecture. Section 4 and 5 present the Hard Real-Time and Soft Real-Time Subsystems of the DEAR-COTS architecture. Finally, Section 6 draws some conclusions and points out possible future directions.

2. RELATED WORK

The problem of reliable real-time DCCS is reasonably well understood when considering synchronous and predictable environments. However, the use of COTS components and the integration,

---
[1] This work was partially supported by the FCT, through project Praxis/P/EEI/14187/1998 (DEAR-COTS).

in the same environment, of hard real-time and soft or non real-time applications introduces new problems.

The guarantee of these properties cannot be achieved by an "ad-hoc" solution, as it has been shown by structured approaches to the problem of designing dependable real-time systems (e.g. DELTA-4 (Powell, 1991), MARS (Kopetz *et al.*, 1989) or GUARDS (Powell *et al.*, 1999)). However, specialised architectures are costly, thus, there is the need for a fault-tolerant real-time distributed architecture for DCCS, based on highly available and low cost COTS technologies.

A few works have addressed the implementation of timeliness requirements in environments where no real-time guarantees can be given (Cristian and Fetzer, 1999; Veríssimo and Almeida, 1995). But the Timely Computing Base model presented in (Veríssimo *et al.*, 2000) provides a generic framework to deal with this problem. It is used as a reference model in the design of the DEAR-COTS architecture. The appropriateness of the TCB model to COTS-based environments is discussed in (Casimiro *et al.*, 2000) for a distributed system composed exclusively by COTS components It concludes that the system must be able to cope with the occurrence of residual timing failures.

Fault tolerance is the preferred means to achieve reliability. A fault-tolerant service can be implemented by co-ordinating a group of processes replicated on different nodes. There are three main replication approaches: active replication, primary-backup (passive) replication and semi-active replication (Powell, 1991). The use of COTS induces fail-uncontrolled replicas, so it becomes necessary the use of active replication techniques. In active replication, all the replicas process the same inputs, keeping their internal state synchronised and voting all on the same outputs.

It is also clear that the feasibility of the applications' requirements must be ensured by sound schedulability analysis techniques. The use of these techniques (e.g., the well-known Response Time Analysis (Audsley *et al.*, 1993)) allows checking if the task set will meet its deadlines, a required condition for hard real-time applications.

3. THE DEAR-COTS ARCHITECTURE

The DEAR-COTS architecture provides an execution environment for real-time applications, with reliability and availability requirements, through the use of COTS hardware and software. Its main purpose is to provide continuous and adequate service to the controlled system, in order to increase the confidence level put in the controlling system. The DEAR-COTS architecture is not targeted to safety-critical systems, as these systems require

a greater level of dependability and a more restricted set of failure assumptions (Laprie, 1992).

Besides the reliability and availability requirements to guarantee the correct behaviour of the supported real-time applications, DCCS also needs to be interconnected with other parts of the overall system. Currently, these kind of systems demand for more flexibility and interconnectivity capabilities, while guaranteeing the availability and reliability requirements of the supported real-time applications. Hence, the integration of hard real-time applications, whose requirements have to be guaranteed, with soft real-time applications, where a more flexible approach can be used, is another goal of the DEAR-COTS architecture.

A common characteristic among all these applications is their real-time behaviour. This real-time behaviour is specified in compliance with timeliness requirements, which in essence call for a synchronous system model. However, the demand for an environment with flexibility and interconnectivity capabilities and based on possibly heterogeneous COTS components, makes the enforcement of timeliness assumptions very difficult. The Timely Computing Base (TCB) model provides a generic solution to this problem.

In the DEAR-COTS architecture the TCB model is used as a reference model to deal with the heterogeneity of system components and of the environment, with respect to timeliness. From a system model perspective we devise a generic DEAR-COTS architecture to address the fundamental problems in a global and integrated way. From an engineering point of view, we devise specific mechanisms to deal with the reliability and availability requirements of hard real-time applications, and to deal with the requirements of soft real-time applications.

3.1 *The TCB model*

The TCB model provides a generic framework to deal with the problem of implementing applications with timeliness requirements in environments that are unpredictable or unreliable. The fundamental idea, which is applied to the DEAR-COTS architecture, is that systems are assumed to have two distinct parts with respect to synchrony. In a system with a TCB there is a *control* part, with synchronous properties, made of local TCB modules interconnected by a *control* channel. There is also a *payload* part, over a global network or *payload* channel, that may have any degree of synchronism. In particular, the payload part can either enjoy synchronous properties but can also be completely asynchronous.

The synchronous part is supposed to be a very small and simple part of the overall system. Therefore, its synchronous properties can be enforced with much higher coverage than if the synchronous properties would have to be assumed for the overall system. Applications execute in the payload part of the system and use a set of basic services provided by the control part, the TCB: *Timely Execution, Duration Measurement* and *Timing Failure Detection*. The definition of these services and the respective API can be found in (Veríssimo *et al.*, 2000).

Applications can benefit from the TCB by construction, using TCB services to "be aware" of their timeliness and, therefore, to behave always in a manner that preserves the safety of the system. This is the case of many soft real-time applications, namely all those that can live with sporadic failures, that can adapt to the available quality of service, or those that can switch to a fail-safe state when some timing failure is detected.

In fact, timing faults affecting components or applications with soft real-time requirements, can be addressed with the help of a timing failure detection service and by applying adequate tolerance or safety measures, as explained in section 5.

3.2 *A generic DEAR-COTS node*

Given the above description of the TCB model, the architecture of a generic DEAR-COTS node, capable of simultaneously handle the requirements of hard and soft real-time applications, can be understood in a very intuitive manner. In fact, the basic idea of casting into the architecture the heterogeneity in system synchrony, has been applied to the generic DEAR-COTS node, as represented by Figure 1.

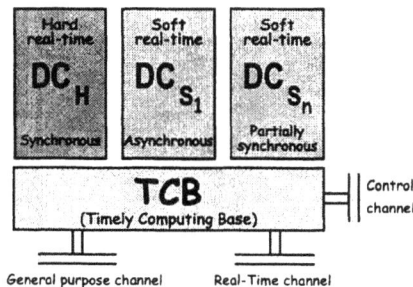

Fig. 1. DEAR-COTS node structure.

The several modules depicted in the figure are intended to fully characterise a node in terms of the synchrony assumptions. The TCB module acts as a gluing element, being always the most synchronous part of the system. It provides timely services to less synchronous modules through an interface that bridges the synchrony gap. The DC_H, DC_{S_1} and DC_{S_n} modules constitute the payload part of the system, and represent the several possible environments with respect to synchrony that may exist in a DEAR-COTS node. Each of these modules can be seen

as a particular subsystem on which applications with different requirements will be executed. DC_H represents a hard real-time subsystem (HRTS), with synchronous properties, which is supposed to exist in every node running distributed and/or replicated hard real-time applications. Different DC_S subsystems can be defined, accordingly to the synchrony properties that they enjoy, namely the asynchronous one, DC_{S_1}, and several partial synchrony models, DC_{S_n} (Cristian and Fetzer, 1999; Veríssimo and Almeida, 1995). With the help of a TCB, all these subsystems (including the asynchronous one) can support the execution of certain applications with timeliness requirements. Therefore, we will refer to all of them as soft real-time subsystems (SRTS).

Each node may have access to different communication channels with respect to synchrony. As in the TCB model, we assume that TCB modules communicate through a control channel, which can be implemented as a virtual channel over the physical (real-time) network or can be a separate network by itself. There is also a real-time channel serving DC_H subsystems, implemented over a real-time network, and a general purpose channel serving DC_S subsystems, where no real-time guarantees are provided.

This generic node architecture can indeed be instantiated into different node configurations, depending on the modules used in a node. A DEAR-COTS system is built by interconnecting several DEAR-COTS nodes, choosing suitable configurations for each node.

3.3 *Implementing a DEAR-COTS system*

A DEAR-COTS system is built using distributed processing nodes, where distributed hard real-time and soft real-time applications may coexist. To ensure the desired level of fault tolerance (reliability and availability) to the supported hard real-time applications, specific components of these applications may be replicated. To ensure the timeliness properties of soft real-time applications and allow system resources to be shared by every (soft and hard real-time) application, processing nodes are modelled as DEAR-COTS nodes.

From a generic DEAR-COTS node it is possible to define particular node instances, with different characteristics and purposes. A DEAR-COTS node is characterised by the synchrony subsystems it is composed of. There are essentially three basic node types: Hard real-time nodes (H), soft real-time nodes (S) and gateway nodes (H/S).

Hard real-time nodes are those where only a hard real-time subsystem (HRTS) exists. Therefore, they will exclusively be used to provide a framework to support reliable and available hard real-time applications, which are the core of the

DCCS application. The existence of a TCB in these nodes is not essential to allow the implementation of hard real-time applications. However, as in any system assumed to be synchronous, and even more in a COTS based one, there is always a small probability that timing failures occur. The need for a TCB, as an instrument to amplify the coverage of synchronous assumptions, has to be equated in terms of the dependability required by the supported DCCS application.

Soft real-time nodes only include a soft real-time subsystem (SRTS). The soft real-time subsystem provides the execution environment for the remote supervision and remote management of the Distributed Computer Control System. The existence of a TCB in these nodes is crucial to allow the implementation of applications with timeliness requirements.

A gateway node integrates both a hard real-time subsystem (HRTS) and a soft real-time subsystem (SRTS). In these nodes there are two distinct and well-defined execution environments. The idea is to allow hard real-time components, executing in the HRTS, to interact in a controlled manner with soft real-time components, executing in the SRTS. The communication mechanisms between both subsystems must be carefully designed, guaranteeing that failures in the soft real-time subsystem (less reliable) do not interfere with the hard real-time subsystem (concerning its timing, availability and reliability requirements). Therefore, mechanisms for memory partitioning must be provided, and also the inter-communication mechanisms must guarantee the integrity of data transferred from the SRTS to the HRTS, by upgrading its confidence level.

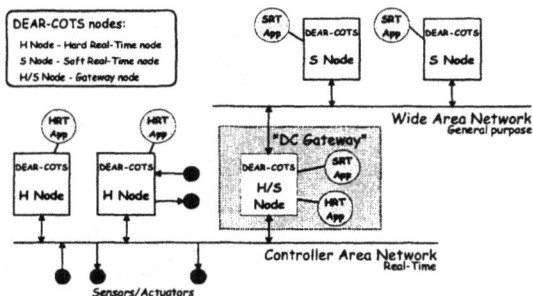

Fig. 2. Generic DEAR-COTS system.

Figure 2 represents a system containing all the above node configurations. H and H/S Nodes are interconnected by a real-time network, which provides the communication infrastructure for the hard real-time applications (interconnecting controllers, sensors and actuators). This real-time network also provides the DEAR-COTS architecture with a communication infrastructure to support the replica management mechanisms. At the above level, as there is the need to interconnect these nodes with the upper levels of the DCCS (e.g. for remote access, remote supervision and/or remote management), there is a general-purpose network interconnecting H/S and S nodes. The control channel required for the TCB is not necessarily an independent network. The assumption of a restricted channel with predictable timing characteristics (control) coexisting with possibly asynchronous channels is feasible in some of the current networks (Prycker, 1995).

At the hard real-time level, the HRTS is responsible for providing a framework for reliable execution. Hence, applications have guaranteed execution resources, including processing power, memory and communication infrastructure. This is the main reason for the need of a separated real-time communication network for the HRTS, where messages sent from one node to another are received and processed in a bounded time interval.

The SRTS provides a set of services to support the supervision and management level of the DCCS. At this system level, flexibility is a major goal, since new services can be provided as the system evolves. However, since there are no hard real-time guarantees, either the TCB services are used to allow applications to be aware of the available quality of service (and adapt themselves, if possible), or techniques such as best-effort scheduling or value-based scheduling must be used.

4. THE HARD REAL-TIME SUBSYSTEM

The set of H and H/S nodes provides a framework to support distributed hard real-time applications (Figure 3). The reliability and availability requirements are guaranteed by the software replication of COTS components, rather than the usual solution, which is to build software on top of specialised hardware.

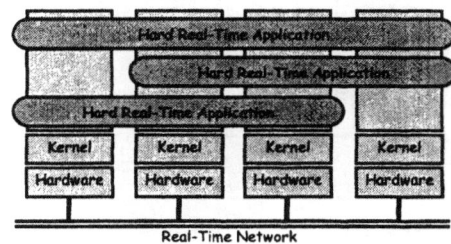

Fig. 3. HRTS structure.

One hard real-time application is constituted by several tasks (processing units). These tasks are distributed over the H and H/S nodes. Each node has its own (non-distributed) COTS real-time kernel and hardware, which provides the desired multitasking support. An advantage of using both a COTS kernel and hardware is that it allows for the easy upgradability and portability of the system.

The set of H and H/S nodes also supports the active replication of software (Figure 4) with dissimilar replicated task sets in each node. By providing different execution environments in each node, the tolerance to design faults is increased, since nodes are considered as independent from the point of view of failures. At the same time flexibility is increased, since nodes are not just copies of each other, allowing for a more flexible design of real-time applications.

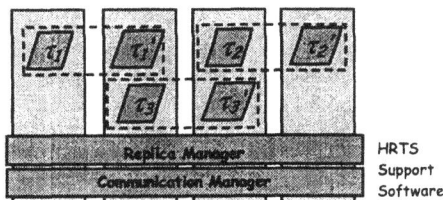

Fig. 4. Replicated Hard Real-Time Application.

The goal of the HRTS support software (Figure 4) is to provide the distribution support (including both the application distribution itself and the replication management) to hard real-time applications. A layered approach to the HRTS is provided to simplify the development of the HRTS support software:

- The Communication Manager layer is responsible for the adequate communication services;
- The Replica Manager layer is responsible for transparently manage the replicated components.

4.1 Scheduling model

In the HRTS, each application consists of a set of related tasks ($\tau_1 \ldots \tau_n$), being each task a single processing unit. Tasks from the same application can be allocated to different nodes. In order to allow the use of current off-line schedulability analysis techniques (e.g., the well-known Response Time Analysis (Audsley et al., 1993)), each task is released only by one invocation event, but can be released an unbounded number of times. A periodic task is released by the runtime (temporal invocation), while a sporadic task can be released either by another task or by the environment. After being released, a task cannot suspend itself or be blocked while accessing remote data (remote blocking).

Tasks are allowed to communicate with each other either through shared data objects or by release event objects (which can also carry data). Shared data objects are used for asynchronous data communication between tasks, while release event objects are used for the release of sporadic tasks. Tasks are designed as small processing units, which, in each invocation, read inputs; carry out the processing; and output the results. The goal is to minimise task interaction, in order to improve the schedulability analysis and increase the system's efficiency. Internal blocking can be bounded

and off-line analysed through the use of Priority Ceiling Protocols (Sha et al., 1990).

4.2 Replication Model

As there is the target of reliability through replication, it is important to define the replication unit. From the above definitions, two different entities could be considered: a single task or the complete hard real-time application. However, both solutions would be very restrictive. In the latter, in order to replicate part of the application, all of its tasks would have to be replicated, which would unnecessarily increase the processing load. In the former, each task output would have to be consolidated, which would increase the inter-task communication load.

Therefore, the notion of *component* is introduced. Applications are divided in components, each one being a set of tasks and resources that interact to perform a common job. The component can include tasks and resources from several nodes, or it can be located in just one node. In each node, several components may coexist. As an example, Figure 5 shows a real-time application with 4 tasks (τ_1, τ_2, τ_3 and τ_4). The application is divided in two different components (C_1 and C_2), which ensure the desired level of reliability.

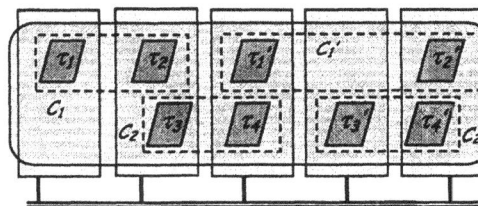

Fig. 5. Hard real-time application.

A similar concept is the "capsule" defined in the Delta-4 architecture (Powell, 1991). As the DEAR-COTS "component", a Delta-4 "capsule" is the unit of replication, embodying a set of tasks (referred as threads) and objects. However, a "capsule" has its own thread scheduling and separated memory space, and is also the unit of distribution. Thus, the Delta-4 concept of "capsule" is more related to Unix processes, whilst the presented component is a more lightweight concept, which is used to structure replication units.

By creating components, it is possible to define the replication degree of specific parts of the real-time application, accordingly to the reliability of its components and the desired level of reliability for the application. However, by replicating components, efficiency decreases as the number of tasks and messages increases. Hence, it is possible to trade reliability for efficiency and vice-versa. Although efficiency should not be regarded as *the* goal of a reliable system, it can be increased by means of decreasing the degree of redundancy of more reliable components.

Nevertheless, as active replication is used, there is the need to guarantee determinism, *i.e.*, that replicated tasks execute with the same data and timing-related decisions are the same in each replica. The use of timed messages (Poledna *et al.*, 2000) allows a restricted model of multitasking to be used and eliminates the need for agreement between the internal tasks of each component. With timed messages, agreement is only needed to guarantee that all replicated components work with the same input values and that they all vote on the final output. The use of timed messages implies the use of clock synchronisation algorithms, since clock deviations must be bounded.

4.3 *HRTS Replica Manager*

The goal of the Replica Manager layer is to provide hard real-time applications with the set of resources required for communication between distributed tasks and between replicated components. In the HRTS, tasks communicate with each other by using shared data and the release of event objects.

If precedence relations exist between tasks, the communication mechanisms can be simplified, since these precedence relations guarantee deterministic execution (Wellings *et al.*, 1998). If the receiving task is sporadic and is released by a sending task, it is guaranteed that, in all replicated components, the replicas of the task will execute with the same data. The same reasoning can be applied when the receiving task is periodic with a period related to the period of the sender task.

The application programmer (transparent approach) does not consider the use of components at the design phase. Later, in a configuration phase, the system engineer configures the components and its replication level, and allocates the different tasks in the distributed system. In this phase, communication streams that need timed messages are identified.

The hindrance of this approach is that, because the programmer is not aware of the possible distribution and replication, complex applications could be built relying heavily in task interaction. However, the model for tasks (presented in Section 4.1), where task interaction is minimised, precludes such applications.

4.4 *HRTS Communication Manager*

The Communication Manager layer is responsible for the adequate communication services, providing a reliable and timely transfer of real-time data.

The group communication abstraction can be used as the framework for reliable communication, and also for replica management (Powell, 1994). In the replication model of DEAR-COTS, a set of replicas from the same component is referred as a group. The goal is to use group communication techniques for simplifying the needed communication mechanisms.

When a task wishes to disseminate its result to more than one task (*1-to-many communication*), reliable multicast algorithms must be used to guarantee that replicated receivers get the same information.

When a task receives inputs from more than one task (*many-to-1 communication*), it must use a consensus algorithm, to choose one of the, possibly different, results it receives, or to compute a result based on the received results. However, as this computation is very application-specific, the application programmer defines appropriate functions for each data stream, which are applied to the set of results.

In *many-to-many communication*, a group of outputs disseminates its results to other group. Each value is the opinion that each member of the group has on the result of the computation. Thus, an interactive consistency algorithm must be used. Once again the programmer must define specific choosing functions.

For *1-to-1 communication* there is no need for specific algorithms besides a reliable transfer of data, as there is no replication.

The suitability of the Controller Area Network (CAN) (ISO, 1993) for the communication infrastructure of the architecture is being studied. Although current results indicate that CAN presents some problems, as it is not resilient to station errors, it is perceived that, with the appropriate fault assumptions, it can be used as the communication infrastructure for the architecture (Pinho *et al.*, 2000).

4.5 *Interconnection with the outside world*

The interconnection of the HRTS with the SRTS in an H/S node must provide mechanisms for transfer of information between both subsystems. These subsystems are in separated memory spaces, in order to prevent errors in the SRTS to interfere with the HRTS behaviour.

Communication from the HRTS to the SRTS does not present major problems, since it is assumed that this information has a higher reliability level. However, if the output to the SRTS comes from replicated components, appropriate agreement must be performed. Conversely, the reliability of the data arriving from the SRTS must be increased, in order to prevent the introduction of erroneous values. As the definition of what is an erroneous data is very application-specific, filters must be defined for each data stream, which must be applied to the incoming data. As before, if the data is to be provided to replicated components, reliable communication algorithms must disseminate this data.

Interconnection with the controlled system is performed through the use of sensors and actuators. Sensor values can be treated as the output of components. The dissemination of the values must be performed using the algorithms identified in the previous section. However, the time at which the value is valid must also be agreed upon.

Output to actuators must also be agreed upon between different replicas. Such agreement may be made either in the computational system or the actuators may perform themselves this agreement, by mechanical or electronic voting on the result. It is out of the scope of the DEAR-COTS project to study actuator agreement whenever such agreement is made outside the computational system.

4.6 *Composition of real-time protocols*

From the point of view of protocol design the run-time must provide a framework to support the clean composition of micro-protocols. This encourages the re-use of protocol components and allows the applications to configure protocol stacks exactly tailored to their needs. This aspect is particularly relevant in the context of real-time applications where, due to memory and power consumption constraints, it is interesting to execute in each component just the protocol layers required to support the intended functionality.

The x-Kernel (Hutchinson and Peterson, 1991) is an early and influential work on protocol composition. A version of x-Kernel adapted to real-time operation has been developed in the scope of the CORDS project (Travostino *et al.*, 1996). Following a similar approach, we are developing *RT-Appia* (Rodrigues *et al.*, 2000), a modern architecture that attempts to balance the flexibility and efficiency of micro-protocols with the predictability requirements of real-time applications.

In *RT-Appia*, a communication channel is an ordered sequence of *sessions*, instances of *protocol layers*. Sessions communicate through the exchange of events. Each channel is executed in the context of a single thread with a fixed and pre-allocated set of memory resources. Layers declare which events need to be subscribed. This ensured that the events are only processed by the relevant layers. Additionally, this knowledge is used to derive the worst case execution time of each channel activation, based on the processing time of each event and on the worst case chain of events that

may be produced in response to a stimuli (from the application, the network or the timer).

5. THE SOFT REAL-TIME SUBSYSTEM

The main goal of the SRTS subsystem is to support the execution of soft real-time applications. The challenging issue is to ensure that applications can have a real-time behaviour despite the occurrence of timing failures.

Timing failures, in a fault-tolerance sense, can be handled either by masking, detection and/or recovery techniques. A generic approach to *timing fault tolerance*, that is, one that can use any or all of the above techniques, requires attributes such as *timeliness*, to act upon failures within a bounded delay, *completeness*, to ensure that failure detection is seen by all participants and *accuracy*, not to detect failures wrongly. In DEAR-COTS, these attributes are mostly ensured by the TCB module and its services.

Separating the mechanisms of timing failure into delay, uncoverage and contamination, allows the introduction of classes of applications that deal with combinations of the former, achieving varying degrees of dependability, when assisted by a TCB: **fail-safe**, which exhibits correct behavior or else stops in fail-safe state; **time-elastic**, which exhibits coverage stability; and **time-safe**, which exhibits no-contamination. In the DEAR-COTS architecture, the idea consists in mapping these application classes (or combinations thereof) into the above-mentioned fault-tolerance techniques. This can be done independently of the synchronism of the SRTS, and allows several degrees of fault tolerance to be achieved, namely:

- Fail-safe operation: by switching to a fail-safe state after the first failure. Requires the timing failure detection service and applications to be of the fail-safe class;
- Reconfiguration and adaptation: by enforcing coverage stability, adapting essential timing variables to environment conditions. Requires applications to be of the time-elastic and time-safe classes.
- Timing error masking: by using replication to mask transient timing errors. Requires accurate timing failure detection and time-safety.

6. CONCLUSIONS

This paper has presented the DEAR-COTS architecture, targeted to the development of reliable distributed computer-controlled systems. It is based in the use of COTS components, and provides a generic framework, in which hard real-time applications can execute, whilst, at the same time, allowing soft or non real-time applications in the system, without interfering with the guarantees provided to hard real-time applications.

DEAR-COTS systems are built using distributed processing nodes, where a mixture of hard real-time and soft real-time applications may execute. The Timely Computing Base model is used as a reference model to deal with the heterogeneity of system components and of the environment with respect to guaranteeing the timeliness of applications.

The paper also describes the main components of the architecture, the Hard Real-Time and Soft Real-Time Subsystems, and its integration in actual implementations of the architecture. The use of the DEAR-COTS architecture allows to fully integrate COTS components and non hard real-time applications with the timeliness and reliability requirements of DCCS.

7. REFERENCES

Audsley, A. N., A. Burns, M. Richardson, K. Tindell and A. Wellings (1993). Applying new scheduling theory to static priority preemptive scheduling. *Software Engineering Journal* 8(5), 285–292.

Casimiro, A., P. Martins and P. Veríssimo (2000). How to build a Timely Computing Base using Real-Time Linux. In: *Proceedings of the 3rd IEEE International Workshop on Factory Communication Systems*. Porto, Portugal. pp. 127–134.

Cristian, F. and C. Fetzer (1999). The timed asynchronous distributed system model. *IEEE Transactions on Parallel and Distributed Systems* 10(6), 642–657. Special Issue on Dependable Real-Time Systems.

Hutchinson, N. and L. Peterson (1991). The x-kernel: An architecture for implementing network protocols. *IEEE Transactions on Software Engeneering* 17(1), 64–76.

ISO (1993). International standard 11898 - road vehicles - interchange of digital information - controller area network (can) for high-speed communication. Technical report.

Kopetz, H., A. Damm, C. Koza, M. Mulazzani, W. Schwabl, C. Senft and R. Zainlinger (1989). Distributed Fault-Tolerant Real-Time Systems: The Mars Approach. *IEEE Micro* 9(1), 25–41.

Laprie, J. C., Ed.) (1992). *Dependability: Basic Concepts and Terminology*. Dependable Computing and Fault Tolerance. Springer-Verlag, Berlin Germany.

Pinho, L. M., F. Vasques and E. Tovar (2000). Integrating inaccessibility in response time analysis of CAN networks. In: *Proceedings of the 3rd IEEE International Workshop on Factory Communication Systems*. Porto, Portugal. pp. 77–84.

Poledna, S., A. Burns, Wellings A. and Barrett P. (2000). Replica determinism and flexible scheduling in hard real-time dependable systems. *IEEE Transactions on Computers* 49(2), 100–111.

Powell, D. (1994). Distributed fault tolerance — lessons learnt from delta-4. *Lecture Notes in Computer Science* 774, 199–217.

Powell, D., Ed.) (1991). *Delta-4 - A Generic Architecture for Dependable Distributed Computing*. ESPRIT Research Reports. Springer Verlag.

Powell, D., J. Arlat, L. Beus-Dukic, A. Bondavalli, P. Coppola, A. Fantechi, E. Jenn, C. Rabéjac and A. Wellings (1999). GUARDS: A Generic Upgradable Architecture for Real-Time Dependable Systems. *IEEE Transactions on Parallel and Distributed Systems* 10(6), 580–599.

Prycker, M. de (1995). *Asynchronous Transfer Mode: Solution For Broadband ISDN*. third ed.. Prentice-Hall.

Rodrigues, J., H. Miranda, J. Ventura and L. Rodrigues (2000). The design of rt-appia. Technical report. Universidade de Lisboa, Faculdade de Ciências.

Sha, L., R. Rajkumar and J. P. Lehoczky (1990). Priority inheritance protocols: An approach to real-time synchronization. *IEEE Transactions on Computers* 39(9), 1175–1185.

Travostino, F., E. Menze and F. Reynolds (1996). Paths: Programming with system resources in support of real-time distributed applications. In: *Proceedings of the 2nd IEEE Workshop on Object-Oriented Real-Time Dependable Systems*. Laguna Beach, CA.

Veríssimo, P., A. Casimiro and C. Fetzer (2000). The Timely Computing Base: Timely actions in the presence of uncertain timeliness. In: *Proceedings of the International Conference on Dependable Systems and Networks*. IEEE Computer Society Press. New York City, USA. pp. 533–542.

Veríssimo, P. and C. Almeida (1995). Quasi-synchronism: a step away from the traditional fault-tolerant real-time system models. *Bulletin of the Technical Committee on Operating Systems and Application Environments (TCOS)* 7(4), 35–39.

Wellings, A., Lj. Beus-Dukic and D. Powell (1998). Real-time scheduling in a generic fault-tolerant architecture. In: *Proceedings of the 19th IEEE Real-Time Systems Symposium*. Madrid, Spain.

Copyright © IFAC Distributed Computer Control Systems,
Sydney, Australia, 2000

AUTOMATED IMPLEMENTATION OF DISTRIBUTED REAL-TIME SYSTEMS USING REAL-TIME OBJECT-ORIENTED MODELING

Saehwa Kim, Sukjae Cho, and Seongsoo Hong

*School of Electrical Engineering and Computer Science
Seoul National University, Seoul, 151-742, Korea*
{ksaehwa, sjcho, sshong}@redwood.snu.ac.kr

Abstract: This paper presents a systematic, schedulability-aware method for automated implementation of complex distributed real-time control systems designed with real-time object-oriented models. Our approach derives tasks in each node of a distributed system by grouping mutually exclusive transactions. It then assigns each task a feasible priority and preemption threshold. To allow for the automated analysis of the timing behavior of the derived task set, our approach exploits a schedulability analysis test based on response time analysis. This paper shows how a run-time system supporting the proposed approach is designed and how code generation is done under our framework. Our method solves the difficult task derivation problem and enables the rapid development of distributed real-time systems. *Copyright ©2000 IFAC*

Keywords: Real-time object-oriented modeling, unified modeling language (UML), real-time scheduling, preemption threshold, multi-threading

1. INTRODUCTION

As distributed real-time control systems get complex and sophisticated to meet the increased degree of safety, reliability, and performance requirements, it becomes inevitable for distributed control system designers to rely on systematic software design methodologies during system development (Boasso, 1993). A wide variety of software design methodologies are available in the market today as commercial products. As is widely accepted, these strategies are classified into two: (1) a task-based design strategy, and (2) an object-oriented design strategy. Object-oriented design methodologies have their prominent advantages over task-based ones since they can allow designers to focus on capturing the high-level abstract features of a system through an object-oriented modeling language capable of expressing high-level system properties. They can also separate the design model of a real-time system from detailed implementation, which in turn allows for easy software maintenance, software reuse, and component-based coding of complex real-time systems that may evolve over time. As object-oriented design methodologies became popular, CASE tools supporting them have emerged, as well.

However, there is an important shortcoming in current object-oriented CASE tools for real-time systems. In object-orient design methodologies, it is not very clear and obvious how to translate an object-oriented design model into an implementation when an underlying run-time system does not support active objects as a first-class entity. This is often the case in practical real-time operating systems since object-based kernels tend to be very slow due to indirect pointers and object refer-

---

[1] The work reported in this paper is supported in part by MOST under the National Research Laboratory (NRL) grant 2000-N-NL-01-C-136 and by Automatic Control Research Center (ACRC).

ences. Thus, object-oriented CASE tools require an additional step of identifying tasks. In practical object-oriented CASE tools, task identification is usually performed in an ad-hoc manner using hints provided by human designers. However, task derivation has a significant effect on the real-time schedulability of the resultant system.

In this paper, we present a systematic method that can automatically generate a multi-threaded implementation from the object-oriented design of a distributed real-time control system. To derive tasks in each node of a distributed real-time system, we rely on the notion of a transaction. It denotes an end-to-end computation from an external input to an external output in each node. It can be described as a sequence of events flowing through the end-to-end computation. Our approach groups mutually exclusive transactions into a task to reduce the number of tasks. It introduces a new task model where each task is a collection of mutually exclusive tasks possessing different scheduling attributes such as periods, execution times, and blocking times. We provide a schedulability analysis algorithm that can take into account this task model. To further reduce the number of tasks, we also adopt preemption threshold scheduling presented in (Wang and Saksena, 1999), and the notion of a non-preemptive group (Saksena and Wang, 2000).

The remainder of this paper is organized as follows. Section 2 overviews the UML-RT design model and its CASE tool. UML-RT is our choice for a real-time object-oriented design model. Section 3 presents a multi-threaded implementation architecture which forms the basis of our approach. Section 4 presents the technical details of our approach. Section 5 explains the implementations of a CASE tool that supports our approach. We conclude this paper in Section 6.

2. OVERVIEW OF UML-RT AND ITS CASE TOOL

In this section, we give a brief overview of UML-RT and RoseRT since UML-RT is used as a source programming language in our approach and RoseRT is the CASE tool of our choice. We also explain their features for distributed system design.

2.1 The UML-RT Modeling Language

UML-RT (Selic and Rumbaugh, 1998) is a specialization of UML (OMG, 1999), and targeted for a category of systems that are characterized as complex, event-driven, and potentially distributed. The new concepts added to UML are derived from the ROOM (Selic *et al.*, 1994) which is also an object-oriented modeling technique optimized for complex, event-driven real-time system.

The fundamental modeling element of UML-RT is a *capsule*. A capsule is a concurrent and possibly distributed object responsible for performing some specific function. A capsule communicates with other objects only by sending and receiving messages through interface objects called *ports*. A capsule may possess internal structures so that it is possible to describe complex systems in a hierarchical manner.

Capsules and connections between capsules' ports construct a structure model. The behavior of each capsule is described by a finite state machine. A capsule's state is changed by a message arrival at a capsule's port. A message can be thought of as an event in that a message has only one method to trigger actions. Thus, we use terms event and message interchangeably. In this paper, we use the UML-RT notations.

UML-RT provides a *deployment diagram* to support distributed system design. A deployment diagram is a graph where each node denotes a processor or a device and an edge between two nodes denotes a connection for point-to-point communication. A processor node represents a computational resource having memory and processing capability and can be further specified with as a collection of components. A component represents the physical package of logical elements such as classes and capsules. A deployment diagram can also be used to model physical elements that may reside on a node, such as executables, libraries, source files, and documents. As a whole, it describes the configuration of processing nodes and their components running on them.

2.2 The RoseRT CASE Tool

RoseRT allows users to design a distributed object-oriented real-time system by providing a visual modeling environment. It also allows them to automatically generate an executable program which is composed of application code, skeleton code, and run-time system service libraries. The core component of the RoseRT run-time system (RTS) is a controller object that dispatches messages to capsule instances. It takes the next message from a message queue and delivers it to the destination capsule instance by invoking the capsule's behavior described by a finite state machine. When there are multiple messages in the queue, it processes them according to their priorities. Thus, message priorities determine the priority of the action triggered by messages.

```
class Message {
    Message*    next;

    Thread      target_thread;
    Capsule     target_object;
    Port        port;

    Short       signal;
    void*       data;

    int         priority;
}
```

Fig. 1. Message data structure.

```
Thread::mainLoop() {
    while(1) {
    wait_for_external_event();
    while(message queues are not empty) {
        foreach(Message M)
            dispatch(M);
        }
    }
}
```
(a)

```
Thread::dispatch(Message M) {
    M.target_object.mutex->enter();
    M.target_object->Behavior(M.signal,M.port);
    M.target_object.mutex->leave();
}
```
(b)

```
An_object::Behavior(int _signal, int _port ) {
    switch ( currentState ) {
    case 0:
        switch( _port ) {
        case 0:
            switch ( _signal ) {
            case 0:
                chain1_t0();
            ...
            }
        ...
        }
    ...
    }
}
```
(c)

Fig. 2. Pseudo code for the main loop of each thread.

RoseRT offers built-in middleware named Connexis which is an off-the-shelf distributed communication infrastructure. It solves many problems common to distributed applications including object-to-object connectivity, fault tolerance, name lookup service, reliability, and performance. Capsules in a distributed system communicate with each other by sending messages to ports just as those in a uniprocessor system do. However, a receiving capsule in a distributed system may be in a different process, or even on a different processor.

3. IMPLEMENTATION ARCHITECTURE

3.1 Implementation Strategies

Depending on implementation strategies, there can be several implementation architectures for an executable program resulted from automatic code generation. The operation of the behavior model in real-time object-oriented modeling is based on event processing. As such, in a multi-threaded implementation architecture, each thread receives its associated messages and processes them one by one. By adopting different strategies for mapping messages (events) to threads, we can come up with as many implementation architectures. As addressed in (Saksena et al., 2000), there can be three different mapping strategies. They are summarized below.

- Map all events associated with an object to a single thread.
- Map all events in a transaction to a single thread.
- Map all events at the same priority to a single thread.

We in turn call them (1) capsule-based, (2) transaction-based, and (3) priority-based mapping. Capsule-based mapping is used in a number of CASE tools including ObjecTime Developer and Rational RoseRT. For example, RoseRT supports a multi-threaded implementation architecture by allowing designers to map capsule in-

stances to multiple tasks. Unfortunately, this requires that the UML-RT design be changed. Designers need to convert their original design model into a dynamically configurable system by associating capsule instances with dynamically created, optional tasks and then inserting code incarnating capsule instances in initial state transitions of fixed capsule instances. This hurts one of the advantages of object-oriented design methods – separation of design models and implementations.

In the next section, we explain transaction-based mapping architecture which forms the basis of our approach. Our approach is based on both the transaction-based and priority-based mapping strategies.

3.2 Transaction-Based Mapping

Each transaction in a system is a sequence of actions flowing through an end-to-end computation initiated by an external event. An action inside a transaction is triggered by an internal message. An external event may initiate a transaction or multiple transactions.

Figure 1 shows the data structure of a message. Note that the message has a reference to a target active object (capsule instance). Figure 2 shows the basic code structure of each thread. Each thread waits for any external event (message) to be delivered. When an external event is delivered, the thread starts a new transaction. The inner while loop of Figure 2 (a) processes an event by invoking an action of a transaction. The run-to-completion semantics of UML-RT requires that the execution of each action be serialized by the mutex guarding an active object possessing the action (Figure 2 (b)). Figure 2 (c) shows the pseudo code of a function (Behavior()) which implements the behavior – represented by a finite state machine – of an active object. This function selects an appropriate action from its active object by matching the port and signal of a delivered message with those of actions, and in turn compares the current state of its active object with the source state of the selected action. If no matched action is found, the delivered message is discarded. Note that the code generator synthesizes the code for each active object.

4. DESIGN MODEL AND PROPOSED APPROACH

In this section, we present our approach that can automatically generate multi-threaded implementations in each node of a distributed real-time system. The source of our approach is a real-time distributed object-oriented design model. Our implementation strategy is the combination of transaction-based mapping and priority-based mapping. Our approach is applied to each node as a three-step process: identifying (1) transactions, (2) logical threads, and then (3) physical threads.

4.1 Description of Design Model

A distributed design model in our approach is composed of a set of processing nodes described in a deployment diagram. Each processing node possesses external and internal events. External events are assumed to include inter-node events so that each node can be processed in isolation. Our approach is based on several assumptions for a distributed design model: (1) All external events are periodic. (2) The deadline of a transaction is the same as its period. This assumption is due to the limitation that end-to-end deadlines cannot be specified in UML-RT. (3) Message priorities are determined not at design time but at implementation time. Our goal is to automate the scheduling of messages through our approach without relying on fine-tuning of a design model. (4) Transactions that start from the same active object are mutually exclusive. This assumption does not impose serious limitations on the utility of our approach since most systems in practice meet this assumption.

We now state our notation. Transaction τ_i denotes a sequence of actions $\langle A_1^i, A_2^i, ..., A_{n_i}^i \rangle$. $O(A_j^i)$ denotes an active object containing action A_j^i. $M(A_j^i)$ denotes a message which triggers action A_j^i. Each action is decomposed into sequence of *sub-actions* $A_j^i = \langle a_{j,1}^i, a_{j,2}^i, ..., a_{j,n_i}^i \rangle$. Sub-action $a_{j,k}^i$ has the worst-case computation time $C(a_{j,k}^i)$. The worst-case computation times of action A_j^i and transaction τ_i can be represented as $\sum_k C(a_{j,k}^i)$ and $C(\tau_i) = \sum_j C(A_j^i)$, respectively. $T(\tau_i)$, $\pi(\tau_i)$, and $\gamma(\tau_i)$ denote the period, priority, and preemption threshold of transaction τ_i, respectively.

4.2 Step 1: Deriving Transactions

Our approach derives a transaction by generating an intermediate tree structure we name a message sequence tree. In a message sequence tree, a node can denote an action and an edge denotes message flow. More specifically, a node denotes either an action or a conjunction or disjunction of messages. Action nodes are classified into *AND-action*, *OR-action*, and *LEAF-action* nodes. An *AND-action* must send out all of its outgoing messages in the left-to-right order. An *OR-action* sends only one of its outgoing messages depending on the condition within the action. A *LEAF-action* does not possess any outgoing messages. When an action has nested conjunctions or disjunctions among its outgoing messages, bridge nodes are used. Similar to action nodes, they are classified into *AND-bridge*, *OR-bridge*, and *LEAF-bridge* nodes. Figure 3 shows an example message sequence tree.

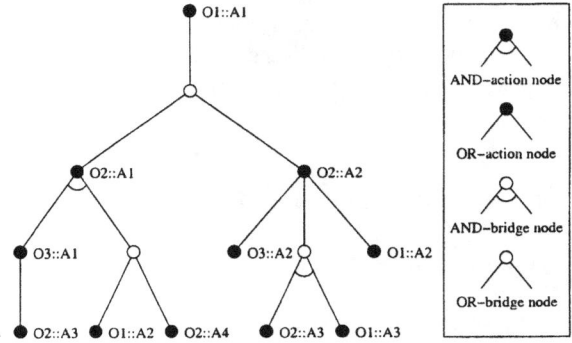

Fig. 3. A message sequence tree.

Our approach derives a transaction in two steps: (1) constructing a tree starting from each action initiated by an external event, and (2) deriving a transaction from each tree.

(1) *Constructing a tree starting from each action initiated by an external event:* The rule to construct a tree is as follows.

- The root node must be an action node.
- The code structure of an action is analyzed so that its path expression – in the form of the regular expression – is generated. In the generated path expression P of an action (node),
  - If P contains no outgoing message. the node becomes a *LEAF* node.
  - Otherwise,
    * If $P = \bigwedge P_i$ $(i = 1, ..., n)$, the node becomes an *AND-action* node having n child nodes for each P_i.
    * If $P = \bigvee P_i$ $(i = 1, ..., n)$, the node becomes an *OR-action* node having n child nodes for each P_i.

* For each child node for a P_i,
  * If $P_i = \bigwedge P_{i,j}$, it becomes an *AND-bridge* node.
  * If $P_i = \bigvee P_{i,j}$, it becomes an *OR-bridge* node.
* The above rule is applied to the nested patterns $P_{i,j,k}, \ldots$ in the same manner.

- If a message triggers different actions, according to the state of the receiving object, an *OR-bridge* node is generated and attached to the node that sends the message.

(2) *Deriving transactions from each tree:* For each message sequence tree, we derive sequences of actions by tracing nodes from the root node. Each sequence of actions becomes a transaction. While tracing nodes,

- When we meet an AND node, we traverse in the depth-first order from left to right for each edge from the node.
- When we meet OR nodes, we generate different sequences for each edge from the node.

In Figure 3, five transactions are derived: $\langle$O1::A1, O2::A1, O3::A1, O2::A3, O1::A2$\rangle$, $\langle$O1::A1, O2::A1, O3::A1, O2::A3, O2::A4$\rangle$, $\langle$O1::A1, O2::A2, O3::A2$\rangle$, $\langle$O1::A1, O2::A2, O2::A3, O1::A3$\rangle$, and $\langle$O1::A1, O2::A2, O1::A2$\rangle$.

4.3 *Step2: Identifying Logical Threads*

Identification of logical threads is a three-step process: for each logical thread, our approach identifies (1) its members, (2) priority, and (3) preemption threshold.

(1) *Grouping transactions:* Identifying the members of a logical thread involves grouping mutually exclusive transactions. Transactions that share an active object as their first action can be grouped into a single logical thread. $L_i = \{\tau_j^i \mid j = 1, 2, \ldots\}$ denotes a logical thread. At run-time, a logical thread may have multiple periods, worst-case execution times, and run-to-completion blocking times since a logical thread may be a collection of transactions.

(2) *Feasible priority assignment:* We adopt Audsley's algorithm (Audsley, 1991) for priority assignment and the response time analysis algorithm (Tindell *et al.*, 1994) for schedulability analysis. We extend the original algorithm to cater for the multiple scheduling attributes of logical threads. The response time $R(L_i)$ of logical thread L_i can be computed using the following equations.

$$\beta(\tau_i) = \max_{\substack{l :: \pi(\tau_l) < \pi(\tau_i), \\ h :: \pi(\tau_h) \le \pi(\tau_i)}} \left\{ \max_{k,m} \{ C(A_k^l) :: O(A_k^l) = O(A_m^h) \} \right\}$$

$$C^R(L_i) = C(\tau_j^i) :: \max_j \{ C(\tau_j^i) + \beta(\tau_j^i) \}$$

$$\beta(L_i) = \beta(\tau_j^i) :: \max_j \{ C(\tau_j^i) + \beta(\tau_j^i) \}$$

$$C^I(L_i) = C(\tau_j^i) :: \max_j \{ C(\tau_j^i) / T(\tau_j^i) \}$$

$$T(L_i) = T(\tau_j^i) :: \max_j \{ C(\tau_j^i) / T(\tau_j^i) \}$$

$$R(L_i) = \beta(L_i) + C^R(L_i) + \sum_{\substack{\forall j, \\ \pi(L_j) > \pi(L_i)}} \left(\left\lceil \frac{R(L_i)}{T(L_j)} \right\rceil \right) \cdot C^I(L_j)$$

where $\beta(\tau_i)$ denotes the blocking time of transaction τ_i.

As explained in Section 3.2, transaction-based implementation architecture should use a mutex lock for each active object to keep the run-to-completion semantics. To reasonably bound the run-to-completion blocking time. we adopt the priority ceiling protocol (L. Sha and Lehoczky, 1990) We assign each mutex a priority ceiling equal to the maximum priority of messages sent to the active object possessing the mutex. With this, only the first action A_1^i of τ_i may be blocked by a lower priority action. For schedulability analysis, we compare response time $R(L_i)$ with deadline $\min_j \{ T(\tau_j^i) \}$.

(3) *Maximum preemption threshold assignment:* Our approach assigns a logical thread the maximum possible preemption threshold since a task with a higher preemption threshold experiences fewer context switches. We use the maximum preemption threshold assignment algorithm and adopt the response time analysis algorithm given in (Saksena and Wang, 2000). Response time $\mathcal{R}(L_i)$ can be computed using the following equations.

$$\beta(\tau_i) = \max_{\substack{l :: \gamma(\tau_l) < \gamma(\tau_i), \\ h :: \gamma(\tau_h) \le \gamma(\tau_i)}} \left\{ \max_{k,m} \{ C(A_k^l) :: O(A_k^l) = O(A_m^h) \} \right\}$$

$$B(L_i) = \max_j \{ C^I(L_i) :: \gamma(L_j) \ge \pi(L_i) > \pi(L_j) \}$$

$$S(L_i) = B(L_i) + \beta(L_i) + \sum_{\substack{\forall j, i \ne j \\ \pi(L_j) \ge \pi(L_i)}} \left(1 + \left\lfloor \frac{S_i}{T(L_j)} \right\rfloor \right) \cdot C^I(L_j)$$

$$\mathcal{R}(L_i) = \mathcal{F}(L_i) = S(L_i) + C^R(L_i) + \\ + \sum_{\forall j, \pi(L_j) > \gamma(L_i)} \left(\left\lceil \frac{\mathcal{F}(L_i)}{T(L_j)} \right\rceil - \left(1 + \left\lfloor \frac{S(L_i)}{T(L_j)} \right\rfloor \right) \right) \cdot C^I(L_j)$$

where $B(L_i)$ is blocking time due to a task with a lower priority and higher preemption threshold, and $S(L_i)$ and $\mathcal{F}(L_i)$ respectively denote the worst-case start time and finish time of logical thread L_i.

With preemption threshold, we assign each mutex a priority ceiling equal to the maximum preemption threshold of messages sent to the active object possessing the mutex. Note that the equation for computing $\beta(\tau_i)$ is changed such that the priority is replaced by the preemption threshold.

4.4 Step3: Identifying Physical Threads

Two logical threads L_i and L_j are mutually non-preemptive if $\pi(L_i) \geq \gamma(L_j)$ and $\pi(L_j) \geq \gamma(L_i)$ (Saksena and Wang, 2000). With this relationship, we can easily construct non-preemptive groups of logical threads where every pair is mutually non-preemptive. We can directly map a non-preemptive group into a physical thread. This can significantly reduce the number of threads, thus run-time context switching overhead and static memory resource demands.

5. TOOL SUPPORT FOR THE PROPOSED APPROACH

While the RoseRT follows the capsule-based mapping strategy, our approach associates each message with an implementation-level thread. Since our approach requires only minimal changes in both the RoseRT run-time system and code generator, it is straightforward to implement it in RoseRT.

Run-time service library:

- We extend the UML-RT message data structure to hold the logical thread's information which includes the priority, the preemption threshold and the physical thread containing the logical thread. We modify the arguments of message sending functions and a mechanism to find a target controller accordingly.

- We change the controller such that it can dynamically determine thread priorities according to the priorities and preemption thresholds of messages in its message queue as in (Saksena and Wang, 2000).

- We use a mutex lock with a priority ceiling for each active object, and modify the message dispatching routine as explained in Section 3.2 and 4.3.

Code generator:

- Our code generator synthesizes the body of `initUpdateThreads()` of the `RTMain` class so that it creates all the derived physical threads. The RoseRT run-time system calls this function once during initialization.

- Our code generator changes calls to the message sending functions to fill their extra arguments so that the attributes of logical threads can be determined via their arguments.

- Our code generator sets values of ceiling priorities on mutexes of active objects.

6. CONCLUSIONS

We have proposed an automatic, schedulability-aware software synthesis method for distributed real-time control systems. The proposed method maps a real-time object-oriented design model to a multi-threaded implementation in each node of a distributed system. It solves the difficult task derivation problem and enables rapid development of distributed control systems designed with real-time object-oriented design modeling.

There are several future research directions. Currently, we are looking to extend the proposed approach to support the notion of a transaction which traverses multiple nodes in distributed real-time control systems.

7. REFERENCES

Audsley, N. (1991). Optimal priority assignment and feasibility of static priority tasks with arbitrary start times. In: *Technical Report YCS 164, Department of Computer Science, University of York, England.*

Boasso, M. (1993). Control systems software. In: *IEEE Transactions on Automatic Contrl.* pp. 1094–1106.

L. Sha, R. Rajkumar and J. Lehoczky (1990). Priority inheritance protocols: An approach to real-time synchronization. *IEEE Transactions on Software Engineering* pp. 1175–1185.

OMG (1999). OMG Unified Modeling Language specification version 1.3.

Saksena, M. and Y. Wang (2000). Scalable real-time system design using preemption thresholds. In: *Submission: Proceedings of IEEE Real-Time Systems Symposium.*

Saksena, M., P. Karvelas and Y. Wang (2000). Automatic synthesis of multi-tasking implementations from real-time object-oriented models. In: *Proceedings of IEEE International Symposium on Object-Oriented Real-Time Distributed Computing.* pp. 360–367.

Selic, B. and J. Rumbaugh (1998). Using UML for modeling complex real-time systems. In: *White Paper, Publicated by ObjecTime, and available from www.objectime.com.*

Selic, B., G. Gullekson and P.T. Ward (1994). *Real-Time Object-Oriented Modeling.* John-Wiley & Sons, Inc.

Tindell, K., A. Burns and A. Wellings (1994). An extendible approach for analyzing fixed priority hard real-time tasks. *Real-Time Systems Journal* pp. 133–151.

Wang, Y. and M. Saksena (1999). Scheduling fixed-priority tasks with preemption threshold. In: *Proceedings of IEEE Real-Time Computing Systems and Applications Symposium.* pp. 328–335.

Copyright © IFAC Distributed Computer Control Systems,
Sydney, Australia, 2000

SEAMLESS INTEGRATION OF REAL-TIME COMMUNICATIONS INTO CAN-CORBA USING EXTENDED IDL AND FAST-TRACK MESSAGES [1]

Gwangil Jeon* Tae-Hyung Kim** Seongsoo Hong***

*Department of Computer Engineering, Seoul National
University, Seoul 151-742, Korea
**Department of Computer Science and Engineering, Hanyang
University, Ansan, Kyunggido 425-791, Korea
***School of Electrical Engineering, Seoul National University,
Seoul 151-742, Korea

Abstract: Component-based middleware technologies have rapidly emerged as a viable solution to the notorious embedded software crisis. However, it is extremely challenging to actually utilize them in real-world applications since embedded systems are subject to a high degree of reliability and real-time requirements, under serious resource requirements. Recently, in (Kim *et al.*, 2000b), we have proposed CAN-CORBA for CAN-based distributed real-time control systems. It addresses major difficulties in designing embedded middleware, such as the narrow bandwidth of the communication medium and the limited size of the payload. In (Jeon *et al.*, 2000), we have also presented a fault-tolerant extension to CAN-CORBA. This paper presents our approach to integrating real-time communications into CAN-CORBA. It first explains how we extend the IDL of the standard CORBA to annotate timing constraints in a CORBA program. It then describes how we can use "fast-track messages" for time-critical real-time traffic. *Copyright ©2000 IFAC*

Keywords: Real-Time Communications, Embedded Systems, Software Engineering, Programming Environments

1. INTRODUCTION

Component-based middleware technologies such as CORBA and DCOM have rapidly emerged as a viable solution to the notorious embedded software crisis. However, it is extremely challenging to actually utilize them in real-world applications since embedded systems are subject to a high degree of reliability and real-time requirements, under serious resource requirements. Recently, in (Kim *et al.*, 2000b), we have proposed an envi-

ronment specific CORBA named CAN-CORBA. It is designed for distributed real-time control systems built on the CAN (Control Area Network) bus (Bosch, 1991). It addresses major difficulties in designing embedded middleware. They include the narrow bandwidth of the CAN bus and the limited payload size of the CAN messages. It also addresses the inappropriate communication model of the original CORBA. As a result of our effort, CAN-CORBA can efficiently run on the CAN bus requiring only small memory footprint and incurring very little run-time overhead.

In (Jeon *et al.*, 2000), we have extended CAN-CORBA to incorporate fault-tolerant features. This extension adopts passive and active repli-

[1] The work reported in this paper was supported in part by MOST under the National Research Laboratory grant, by KOSEF under grant 981-0924-127-2, and by Automatic Control Research Center (ACRC).

cation strategies mandated by the OMG fault-tolerant CORBA draft standard, and it makes use of a state-less passive replication policy to reduce resource demands of these fault-tolerant features.

This paper presents a real-time extension to the CAN-CORBA – the final missing link for embedded middleware. We aim at seamlessly integrating real-time communications into CAN-CORBA with only minimal changes to the original CAN-CORBA design. Our extension includes two ingredients. First, we extend the standard IDL (Interface Description Language) of CORBA to allow programmers to annotate timing constraints in a CORBA program. Second, we define "fast-track messages" and use them for time-critical real-time traffic in a CAN bus. A fast-track message is a special class message that can be transmitted via the CAN bus with only a minimum amount of interference from non-real-time traffic. When an IDL program contains timing constraints, fast-track messages are generated. They can bypass the entire layers of abstraction in the CAN-CORBA software bus and be sent with priority using the priority-based bus arbitration mechanism of the CAN. These fast-track messages allow the real-time CAN-CORBA to guarantee timely delivery of real-time messages according to their deadlines and priorities. If an IDL program contains no timing constraints, it is simply processed by a normal IDL compiler as usual. In this way, we are able to seamlessly integrate real-time communications into the CAN-CORBA.

The remainder of the paper is organized as follows. Section 2 gives the target system hardware model and the transport protocols for CAN-CORBA. In Section 3, we define timing constraints and how to specify these constraints in a IDL program. We present how to generate the fast-track CAN messages from the extended IDL specification in Section 4. Finally, Section 5 concludes this paper.

2. SYSTEM MODEL

CAN-CORBA is designed to operate on a distributed embedded system built on the CAN bus. In this section, we present the target system hardware model to help readers understand the characteristics and limitations of the underlying system platform, before delving into the details of the real-time CAN-CORBA design.

2.1 *Target System Hardware Model*

The target hardware, which is same as in our previous work (Jeon *et al.*, 2000), (Kim *et al.*, 2000*a*), and (Kim *et al.*, 2000*b*), consists of a number of function control units (FCU) interconnected by

embedded control networks (ECN). As an example of the model, Figure 1 shows the electronic control system of a passenger vehicle. Each FCU, possessing one or more microcontrollers and microprocessors, conducts a dedicated control mission by interfacing sensors and actuators and executing prescribed control algorithms. Depending on configuration, an FCU works as a data producer, a consumer, or both.

As shown in Figure 1, embedded control networks (ECN) connect FCUs through inexpensive bus adaptors. Such ECNs are often required to provide real-time message delivery services and subject to very stringent operational and functional constraints. In this work, we have chosen the CAN (ISO-IS 11898, 1993) as our embedded control network substrate since it is an internationally accepted industrial standard satisfying such constraints.

The CAN is well suited for real-time communication since it is capable of bounding message transfer latencies via predictable, priority-based bus arbitration. A CAN message is composed of identifier, data, error, acknowledgment, and CRC fields. The identifier field consists of 11 bits in CAN 2.0A or 29 bits in 2.0B and the data field can grow up to eight bytes. When a CAN network adaptor transmits a message, it first transmits the identifier followed by the data. The identifier of a message serves as a priority, and a higher priority message always beats a lower priority one.

The CAN provides a unique addressing scheme, known as subject-based addressing (Oki *et al.*, 1993). In the CAN, a message put into the network does not contain its destination address. Instead, it contains a subject tag – a predefined bit pattern in the message identifier which serves as a hint about its data content. A receiver node can program its CAN bus adaptor to accept only a specific subset of messages that carry a specific identifier pattern with them. This filtering mechanism is made possible via a mask register and a set of comparison registers on a CAN interface chip. This subject-based addressing scheme is a key underlying mechanism for the communication models of our CAN-CORBA.

2.2 *CAN-CORBA Communication Channels*

CAN-CORBA offers a subscription-based, anonymous group communication scheme that is often referred to as "blindcast" or as a publisher/subscriber scheme (Rajkumar *et al.*, 1995), (Kaiser and Mock, 1999). In this scheme, a communication session starts when a data producer announces a predefined invocation channel. An invocation channel is a virtual broadcast channel

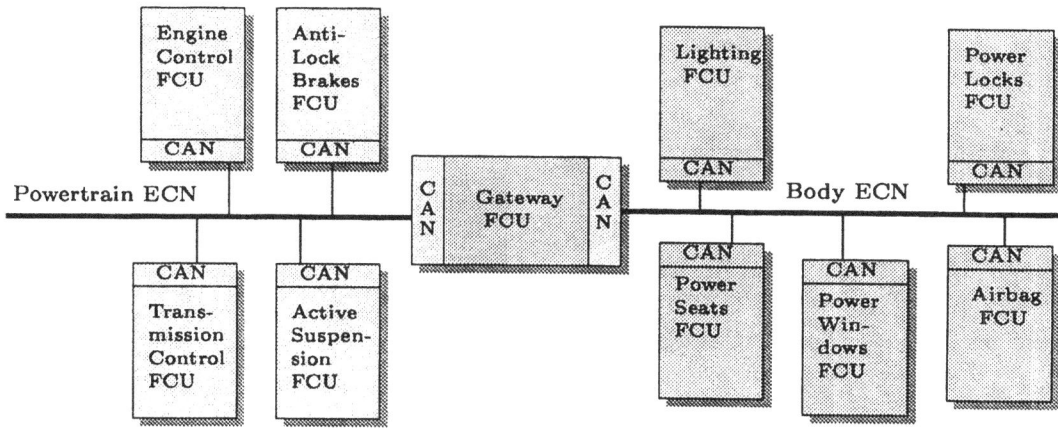

Fig. 1. Example distributed embedded control system: Passenger vehicle control system.

from publishers to a group of subscribers. Data consumers can subscribe to an announced invocation channel. In this announcement/subscription process, neither a publisher nor a subscriber has to know each other. This anonymity allows for easy reconfiguration of control systems. In CAN-CORBA, an invocation channel is uniquely identified with a CAN identifier, and maintained by the *conjoiner* as described in Section 2.3.

CAN-CORBA also provides point-to-point communication primitives for interoperability with other standard CORBA implementations. Since the CAN bus is a broadcast medium, the publisher/subscriber model is more natural and efficient than the point-to-point model. Moreover, fault tolerance is better serviced by a group communication scheme. Readers are referred to our previous work in (Kim *et al.*, 2000*b*) for more details on the connection-oriented communications in CAN-CORBA.

2.3 *Transport Layer Protocol of CAN-CORBA*

Since the CAN standard specifies physical and data link layer protocols in the OSI reference model (Bosch, 1991), we have developed our own transport protocol for the CAN-CORBA. Figure 2 shows the protocol header format. We divide the CAN identifier structure into three sub-

fields: a protocol ID (Proto), a transmitting node address (TxNode), and a port number (TxPort). They respectively occupy two, five and four bits amounting to 11. The Proto field denotes an upper layer protocol identifier. The upper layer protocols include the publisher/subscriber (01), the point-to-point (10), and the binding protocol (11). We deliverately choose protocol ID (00) for real-time messages since it is the highest prioirty message class in the CAN bus context. Note that a message identifier with a smaller value gets a higher priority during bus arbitration in the CAN. When a CAN protocol header denotes a real-time message, the remaining 9 bits are used to contain the message identifer just as normal CAN messages do. In other words, TxNode and TxPort do no longer mean for node address and port number, but are used to determine the urgency of the message according to its deadline and priority.

2.4 *Fast-track Vs. CAN-CORBA Messages*

While the fast-track messages and the CAN-CORBA messages are represented in the same CAN message format, they takes the totally different path to be transmitted in the software bus hierarchy. The former merely relies on the CAN bus arbitration mechanism. The latter messages are intercepted and interpreted appropriately by the corresponding CAN-CORBA objects for inter-object communication that is supported by the CAN-CORBA ORB (Object Request Broker). In short, the fast-track messages are *CAN messages*, and the CAN-CORBA messages are the *CORBA messages* that contain the "middleware" implications.

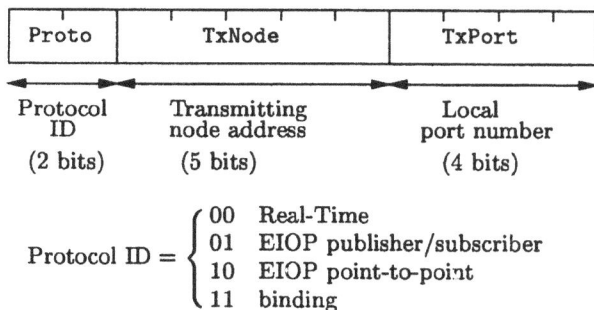

Fig. 2. Protocol header format using CAN identifier structure

3. EXTENDED IDL FOR TIMING CONSTRAINTS

Object interconnection activity is understood to be an essentially distinct and different intellectual activity from that of implementing individual objects. Unlike the module implementation languages such as C++ and Java, the IDL is the language used to describe the interfaces that client objects call and object implementations provide (OMG, 1998). Therefore, an interface definition written in IDL completely defines the inter-object communication without having to concern about implementation details.

Since a component in a distributed control system can be specified in a task-level end-to-end timing constraints, it is quite adequate to extend IDL to include timing constraints as one of inter-object behavior specifications. It is a better approach than embedding timing constraints in a module implementation language, not only because the IDL level modification is much simpler than writing a new compiler, but because the clear distinction between the function and the behavior part of a task helps to write and revise a control program much easier. Whenever timing constraints are to be changed, programmers need to revise an IDL definition only leaving source codes unchanged. In this section, we present how to specify timing constraints in the so-called extended IDL (*xIDL*), and how to develop an application with and without real-time tasks using xIDL and "fast-track" messages in CAN-CORBA.

3.1 *Real-Time Application Model*

We presume that a real-time distributed control application is composed of a *static* set of hard real-time messages (typically control messages), and each message has a statically assigned priority. Each real-time task is also statically assigned to a CAN node station. A recipient task is invoked by an interrupt handler in CAN device controller upon receiving a message after filtering irrelevant messages. A sender task periodically broadcasts control messages of higher priorities over non-real-time messages in the form of fast-track messages. Tindell and Burns claimed that we can bound the worst-case response time of a hard real-time message if each message has a bounded size and a bounded rate (Tindell and Burns, 1994).

We also assume a programmer obtained a feasible schedule for a set of periodic messages before runtime with an aid of schedulability check tools or whatsoever fixed priority scheduling methods, and all necessary real-time factors are available at the time of writing a distributed control program. Dynamic scheduling method with sporadic and aperiodic messages will not be covered in this paper.

Although we do not allow communications between real-time and non-real-time objects in an

```
interface TemperatureMonitor {
  oneway void update_temperature(in char
    locationID, in long temperature)
    with (period = 200, deadline = 150,
      release_time = 10, priority = 15)
    raises (DeadlineMissed);
}
```

Fig. 3. xIDL definition for a real-time method.

application, the entire CAN bus system can be shared with other non-real-time distributed control applications.

3.2 *Extended IDL Specification*

The xIDL is extended to include a **with** clause that specifies timing constraints such as priority, period, deadline and release time. An interface possesses a set of methods and attributes. Note that it is possible for an interface to have some non-real-time (i.e. ordinary CORBA) methods and real-time methods Only real-time methods are liable for timing constraints. Thus, the existence of **with** clauses as shown in Figure 3, where four timing constraints are specified for a real-time method, makes the only difference between IDL and xIDL.

In the figure, the interface **TemperatureMonitor** has a real-time method update_temperature which emits a control message of priority[2] 15 at every 200 milliseconds, no later than 150 milliseconds, no earlier than 10 milliseconds. If one of the constraints is violated, it raises an exception. In summary, if a method has a **with** phrase, it is a real-time method, and the information is sifted out by the xIDL preprocessor and delivered to the real-time task generator to make a source-level transformation as illustrated in Figure 4.

3.3 *Real-time Application Program Development Process*

Figure 4 illustrates the entire process of application program generation starting at the xIDL specification level in the real-time CAN-CORBA.

If an xIDL definition contains no timing constraint (i.e. same as a normal IDL), the generation path exactly follows the same one in an ordinary CAN-CORBA application case as stated in (Kim *et al.*, 2000a), (Kim *et al.*, 2000b), and (Jeon *et*

---

[2] CAN-CORBA can specify up to 512 (that is, 2^9) priorities on the whole. So the priority has a fixed range of [0..511]. Recall that CAN 2.0A identifier field consists of 11 bits and the two most significant bits are reserved to denote a protocol ID.

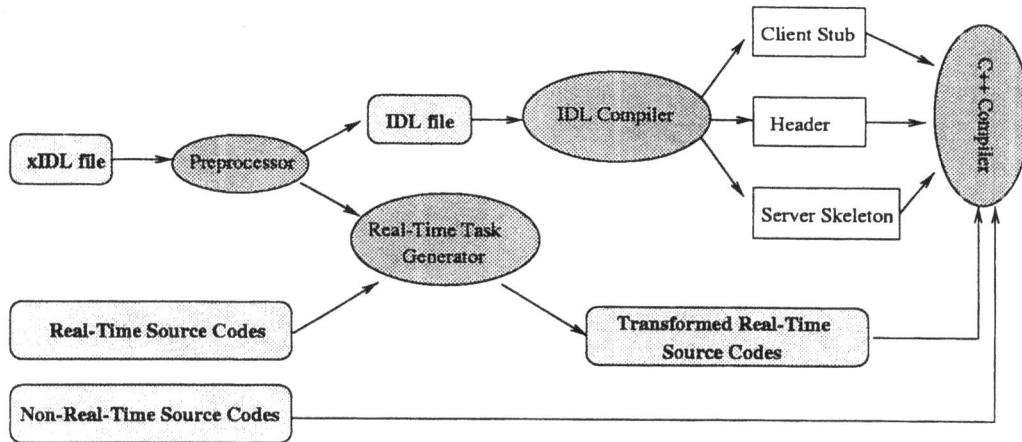

Fig. 4. Application program development process from xIDL in the real-time CAN-CORBA.

al., 2000). However, if it does, the preprocessor sifts out the timing constaint from an xIDL definition by reading the file. As a result, the ordinary IDL is generated for the IDL compiler, and the sieved timing constraint is used to generate the "fast-track" messages according to the fixed priority scheduling method. The real-time task generator actually performs a source-code level transformation; the interface source code, which is involved in an xIDL definition with a timing constraint, is transformed to the one that generates "fast-track" messages for the CAN bus.

For instance, suppose there is an object that broadcasts a sensed temperature message. Also suppose there is an interface definition with a timing constraint for the object in xIDL, i.e., that mandates a periodic delivery of the message at a given rate. Then the object must be a real-time task, and the messages must precede all other non-real-time CAN-CORBA messages. Moreover, those messages among many real-time tasks must be sorted by priorities according to a given fixed priority schedule. To this end, the source code that implements the object will be appropriately transformed to guarantee the timely delivery of the messages by the real-time task generator based on the timing constraint in the xIDL. In this way, the real-time communications can be seamlessly integrated to the CAN-CORBA, by keeping the application programmer's intervention minimal – only at the xIDL level.

4. REAL-TIME TASK GENERATION

In this section, we present the role of the "Real-Time Task Generator" in Figure 4. It transforms real-time source codes so that the resulting codes contain fast-track messages for timely delivery of real-time messages according to the timing constraints in xIDL. We describe how to generate such a real-time task for a general specification of

timing constraint: with (priority = p, period = t, deadline = d, release_time = r).

4.1 Message Identifier Preparation for Fast-Track Message using p

A message identifier determines the priority of a message in the CAN bus. The predetermined priority p, possibly through a fixed priority scheduling, is directly used in preparing the corresponding fast-track message.

Hard real-time control messages should have priority over non-real time messages to acquire the CAN bus. The 11-bit message identifier determines the priority: the smallest value has the highest priority. The CAN bus arbitrates among many conflicting messages according to the ascending order of their identifiers. As explained in Section 2, the protocol id (Proto) for fast-track messages is 00_2. Therefore, the messages have always higher priority over other messages. Note that a priority in a with phrase has a fixed range of $[0..511]$. Thus, the message identifier for a particular real-time object has a concatenation of 00_2 and the 9-bit binary representation of p.

4.2 Operational Code Generation for t, d, and r

A real-time source code written by a programmer in Figure 4 is merely a function body for a task that are free from explicit time-related statements. It is our intention to separate the real-time factors from the object implementation level. Those factors are determined a priori through a separate real-time design process, and subject to change if the previous design becomes to raise an exception of DeadlineMissed(). Since the factors are expressed in the interface definition language level, programmers do not have to reinvent the wheel manually simply because of the frequent changes in their values.

131

```
int k = 1;
while (1) {
        start = gettime();
        // calculate the message contents
        msg = ReadTemperature();
        // keep the release time spec.
        while (gettime() < start + r) ;
        send(msg);
        if (gettime() > t + d)
                DeadlineMissed();
        while (gettime() <= k * t); k++
}
```

Fig. 5. Example: a transformed real-time source code.

The real-time task generator accepts the source code and timing constraints t, d, r and generates a transformed source code that implements the timing specification. Figure 5 shows an example of the transformed code with respect to t, d, r. ReadTemperature() is the functional body that is written by a programmer. All other codes are automatically generated from the timing constraints in an xIDL definition.

Given a bounded time, the ReadTemperature() function completes the task of reading temperature. The msg is now ready to send but not before the specified release time r. Then, the message is sent via the CAN bus by calling send(). Since the worst-case queuing time (at a sender-side buffer in the bus) and transmission delay are bounded, the physical message is guaranteed to be delivered in a bounded time. The deadline checkup is an assertive statement for defensive programming because an exception should not be raised if the design is correct. Finally, the next cycle is started after p.

5. CONCLUSIONS

Developing an embedded software is a complicated, time-consuming, and error-prone task. A useful software architectural support is indispensable for embedded system engineers to cope with the embedded software crisis. We have developed an environment specific CORBA for CAN-based distributed real-time control systems. In this paper, we have presented how to integrate real-time communications into CAN-CORBA seamlessly, which is the final missing link in a series of our research effort (Kim et al., 2000a), (Kim et al., 2000b), (Jeon et al., 2000) for a better distrbited control programming environment.

In order to guarantee the timing performance of hard real-time control messages, these messages are transmitted as fast-track messages that penetrate entire layers in the CAN-CORBA software bus. Real-time CAN-CORBA applications are described in xIDL which has an extension of a with phrase to specify timing constraints. The task generator automatically generates real-time source codes using the timing annotations in xIDL. Since all timing constraints are expressed in xIDL, and the relevant source codes are automatically generated, programmers are free to experiment various combinations of timing factors without having to worry about rewrite their source codes directly.

6. REFERENCES

Bosch (1991). CAN specification, version 2.0.

ISO-IS 11898 (1993). Road vehicles - interchange of digital information - controller area network (CAN) for high speed communication.

Jeon, G., T. Kim, S. Hong and S. Kim (2000). A fault tolerance extension to the embedded corba for the can bus systems. In: *ACM SIGPLAN 2000 Workshop on Languages, Compilers, and Tools for Embedded Systems.*

Kaiser, J. and M. Mock (1999). Implementing the real-time publisher/subscriber model on the controller area network (CAN). In: *IEEE International Symposium on Object-oriented Real-time distributed Computing.*

Kim, K., G. Jeon, S. Hong, S. Kim and T. Kim (2000a). Resource conscious customization of CORBA for CAN-based distributed embedded systems. In: *IEEE International Symposium on Object-Oriented Real-Time Computing.*

Kim, K., G. Jeon, S. Hong, T. Kim and S. Kim (2000b). Integrating subscription-based and connection-oriented communications into the embedded corba for the can bus. In: *IEEE Real-time Technology and Application Symposium.*

Oki, B., M. Pfluegl, A. Seigel and D. Skeen (1993). The information bus – an architecture for extensible distributed systems. In: *ACM 14th Symposium on Operating System Principles.*

OMG (1998). The Common Object Request Broker: Architecture and specification revision 2.2.

Rajkumar, R., M. Gagliardi and L. Sha (1995). The real-time publisher/subscriber interprocess communication model for distributed real-time systems: Design and implementation. In: *IEEE Real-time Technology and Application Symposium.*

Tindell, K. and A. Burns (1994). Guaranteeing message latencies on controller area network (can). In: *Proceedings of the International CAN Conference.*

AUTHOR INDEX

www.ingramcontent.com/pod-product-compliance
Lightning Source LLC
Chambersburg PA
CBHW082307210326

41598CB00028B/4461